2020: An Astronomical Year

A Reference Guide to 365 Nights of Astronomy

Richard J. Bartlett

Top Cover Image: This image shows our neighbouring planet Mars, as it was observed shortly before opposition in 2016 by the NASA/ESA Hubble Space Telescope. Some prominent features of the planet are clearly visible: the ancient and inactive shield volcano Syrtis Major; the bright and oval Hellas Planitia basin; the heavily eroded Arabia Terra in the centre of the image; the dark features of Sinus Sabaeous and Sinus Meridiani along the equator; and the small southern polar cap.

Mars reaches opposition again on October 14th this year. It will be the best opposition of the planet until 2035.

Credit: NASA, ESA, the Hubble Heritage Team (STScI/AURA), J. Bell (ASU), and M. Wolff (Space Science Institute)

Image source: https://commons.wikimedia.org/wiki/File:Mars_in_opposition_2016.jpg

Bottom Cover Image: Valles Marineris is a vast canyon system that runs along the Martian equator. At 4,500 km long, 200 km wide and 11 km deep, the Valles Marineris rift system is ten times longer, seven times wider and seven times deeper than the Grand Canyon of Arizona, making it the largest known crevice in the solar system.

Credit: NASA / JPL-Caltech / USGS

Image source: https://commons.wikimedia.org/wiki/File:VallesMarinerisHuge.jpg

First Edition, August 2019

Contents

Introduction

About the 2020 Edition

I'm very pleased to say that, for the first time, I've been able to include depictions of night sky events in the paperback edition of this book. It's something I've tried to incorporate in the past, but unfortunately, I was never able to get the images to look the way I wanted. (And printing in color was simply too expensive.)

So now, thanks to the *SkySafari* app (available on both Apple and Android phones) and the folks at Simulation Curriculum Corp., I've finally been able to include those images. Incidentally (and I haven't been paid to say this) I have to say that *SkySafari* has proven to be very useful!

I used Kansas, KS as the basis for the images; I tried to pick a location that was as central to North America as possible. Obviously, the view will differ a little from your own location, but I've found that the depictions still work well regardless of where you are. (For example, I live in Los Angeles and can't say the views have been radically different.)

As always, I'm indebted to Wolfgang Zima and his *Mobile Observatory* app for Android. I've used *MO* for the graphical depictions of the planets, as well as the charts depicting their positions and the star charts at the beginning of the book.

I'm always looking for ways to improve the book and am open to comments and suggestions. With that in mind, please feel free to email me at astronomywriter@gmail.com if there's anything you'd like to see included.

About the Author

Photo by my son, James Bartlett

I've had an interest in astronomy since I was six and although my interest has waxed and waned like the Moon, I've always felt compelled to stop and stare at the stars.

In the late 90's, I discovered the booming frontier of the internet, and like a settler in the Midwest, I quickly staked my claim on it. I started to build a (now-defunct) website called *StarLore*. It was designed to be an online resource for amateur astronomers who wanted to know more about the constellations - and all the stars and deep sky objects to be found within them. It was quite an undertaking.

After the website was featured in the February 2001 edition of *Sky & Telescope* magazine, I began reviewing astronomical websites and software for their rival, *Astronomy*. This was something of a dream come true; I'd been reading the magazine since I was a kid and now my name was regularly appearing in it.

Unfortunately, a financial downturn forced my monthly column to be cut after a few years but I'll always be grateful for the chance to write for the world's best-selling astronomy magazine.

I emigrated from England to the United States in 2004 and spent three years under relatively clear, dark skies in Oklahoma. I then relocated to Kentucky in 2008 and then California in 2013. I now live in the suburbs of Los Angeles; not the most ideal location for astronomy, but there are still a number of naked eye events that are easily visible on any given night.

The Author Online

Amazon US: http://tinyurl.com/rjbamazon-us   Amazon UK: http://tinyurl.com/rjbamazon-uk

Facebook: http://tinyurl.com/rjbfacebook   Twitter: http://tinyurl.com/rjbtwitter

Email: astronomywriter@gmail.com

Clear skies,

Richard J. Bartlett

August 22nd, 2019

Star Charts & Observing Lists

Star Chart Tables

If observing during daylight savings time, first deduct one hour and then refer to the corresponding chart number. For example, for 10pm daylight savings time in early August, use chart 18.

	6pm	7pm	8pm	9pm	10pm	11pm
Early January	1	2	3	4	5	6
Late January	2	3	4	5	6	7
Early February	3	4	5	6	7	8
Late February	4	5	6	7	8	9
Early March	5	6	7	8	9	10
Late March	6	7	8	9	10	11
Early April	7	8	9	10	11	12
Late April	8	9	10	11	12	13
Early May	9	10	11	12	13	14
Late May	10	11	12	13	14	15
Early June	11	12	13	14	15	16
Late June	12	13	14	15	16	17
Early July	13	14	15	16	17	18
Late July	14	15	16	17	18	19
Early August	15	16	17	18	19	20
Late August	16	17	18	19	20	21
Early September	17	18	19	20	21	22
Late September	18	19	20	21	22	23
Early October	19	20	21	22	23	24
Late October	20	21	22	23	24	1
Early November	21	22	23	24	1	2
Late November	22	23	24	1	2	3
Early December	23	24	1	2	3	4
Late December	24	1	2	3	4	5

If observing during daylight savings time, first deduct one hour and then refer to the corresponding chart number. For example, for 2am daylight savings time in early July, use chart 20.

	12am	1am	2am	3am	4am	5am	6am
Early January	7	8	9	10	11	12	13
Late January	8	9	10	11	12	13	14
Early February	9	10	11	12	13	14	15
Late February	10	11	12	13	14	15	16
Early March	11	12	13	14	15	16	17
Late March	12	13	14	15	16	17	18
Early April	13	14	15	16	17	18	19
Late April	14	15	16	17	18	19	20
Early May	15	16	17	18	19	20	21
Late May	16	17	18	19	20	21	22
Early June	17	18	19	20	21	22	23
Late June	18	19	20	21	22	23	24
Early July	19	20	21	22	23	24	1
Late July	20	21	22	23	24	1	2
Early August	21	22	23	24	1	2	3
Late August	22	23	24	1	2	3	4
Early September	23	24	1	2	3	4	5
Late September	24	1	2	3	4	5	6
Early October	1	2	3	4	5	6	7
Late October	2	3	4	5	6	7	8
Early November	3	4	5	6	7	8	9
Late November	4	5	6	7	8	9	10
Early December	5	6	7	8	9	10	11
Late December	6	7	8	9	10	11	12

Chart 1

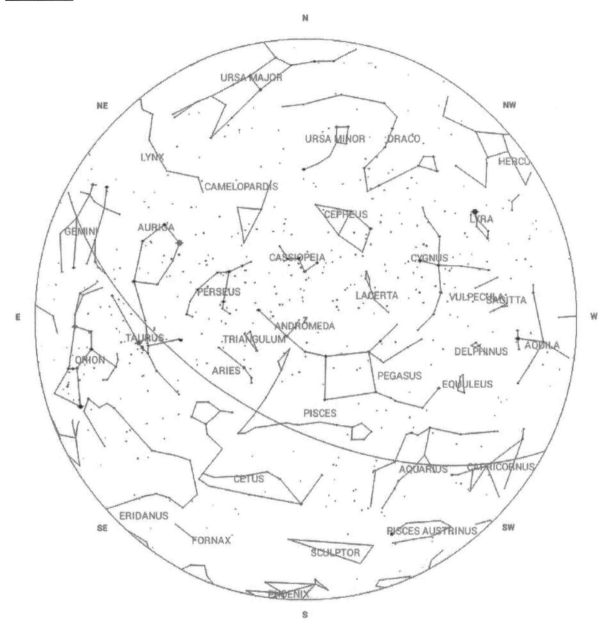

Designation	Name	Con.	Type	R.A.	Dec.	Mag	Size/Sep
Gam And	Almach	And	MS	02h 04m	+42° 20'	2.1	10"
M 31	Andromeda Galaxy	And	Gx	00h 43m	+41° 16'	4.3	156'
NGC 752	Golf Ball Cluster	And	OC	01h 58m	+37° 47'	6.6	75'
Pi And		And	MS	00h 38m	+33° 49'	4.4	36"
107 Aqr		Aqr	MS	23h 47m	-18° 35'	5.3	7"
94 Aqr		Aqr	MS	23h 19m	-13° 28'	5.2	13"
NGC 7293	Helix Nebula	Aqr	PN	22h 30m	-20° 50'	6.3	16'
Zet Aqr		Aqr	MS	22h 29m	-00° 01'	3.7	2"

Designation	Name	Con.	Type	R.A.	Dec.	Mag	Size/Sep
41 Aqr		Aqr	MS	22h 14m	-21° 04'	5.3	5"
53 Aqr		Aqr	MS	22h 27m	-16° 45'	5.6	3"
Lam Ari		Ari	MS	01h 59m	+23° 41'	4.8	37"
Gam Ari	Mesarthim	Ari	MS	01h 54m	+19° 22'	4.6	8"
30 Ari		Ari	MS	02h 37m	+24° 39'	6.5	39"
Kemble 1	Kemble's Cascade	Cam	Ast	03h 57m	+63° 04'	5.0	180'
Sig Cas		Cas	MS	23h 59m	+55° 45'	4.9	3"
NGC 457	Owl Cluster	Cas	OC	01h 20m	+58° 17'	5.1	20'
M 52		Cas	OC	23h 25m	+61° 36'	8.2	15'
Struve 163		Cas	MS	01h 51m	+64° 51'	6.5	35"
Struve 3053		Cas	MS	00h 03m	+66° 06'	5.9	15"
Eta Cas	Achird	Cas	MS	00h 50m	+57° 54'	3.6	13"
M 103		Cas	OC	01h 33m	+60° 39'	6.9	5'
NGC 281		Cas	OC	00h 53m	+56° 38'	7.4	4'
Iot Cas		Cas	MS	02h 29m	+67° 24'	4.5	7"
NGC 7789	Herschel's Spiral Cluster	Cas	OC	23h 57m	+56° 43'	7.5	25'
NGC 559		Cas	OC	01h 30m	+63° 18'	7.4	6'
NGC 659	Ying Yang Cluster	Cas	OC	01h 44m	+60° 40'	7.2	5'
NGC 663		Cas	OC	01h 46m	+61° 14'	6.4	14'
Del Cep		Cep	MS/Var	22h 29m	+58° 25'	3.5-4.4	41"
Xi Cep	Alkurhah	Cep	MS	22h 04m	+64° 38'	4.3	8"
Omi Cep		Cep	MS	23h 19m	+68° 07'	4.8	3"
Gam Cet	Kaffajidhma	Cet	MS	02h 43m	+03° 14'	3.5	3"
Omi Cet	Mira	Cet	Var	02h 19m	-02° 59'	2.0-10.1	N/A
NGC 7243		Lac	OC	22h 15m	+49° 54'	6.7	29'
8 Lac		Lac	MS	22h 36m	+39° 38'	5.7	82"
Bet Per	Algol	Per	Var	03h 08m	+40° 57'	2.1-3.4	N/A
Eps Per		Per	MS	03h 58m	+40° 01'	2.9	9"
M 34		Per	OC	02h 42m	+42° 46'	5.8	35'
NGC 869/884	Double Cluster	Per	OC	02h 21m	+57° 08'	4.4	18'
Eta Per		Per	MS	02h 51m	+55° 54'	3.8	29"
NGC 1245		Per	OC	03h 15m	+47° 15'	7.7	10'
Melotte 20	Alpha Persei Moving Cluster	Per	OC	03h 24m	+49° 52'	2.3	300'
TX Psc		Psc	Var/CS	23h 46m	+03° 29'	4.5-5.3	N/A
Alp Psc	Alrisha	Psc	MS	02h 02m	+02° 46'	3.8	2"
Psi1 Psc		Psc	MS	01h 06m	+21° 28'	5.3	30"
Zet Psc		Psc	MS	01h 14m	+07° 35'	5.2	23"
55 Psc		Psc	MS	00h 40m	+21° 26'	5.4	6"
65 Psc		Psc	MS	00h 50m	+27° 43'	7.0	4"
M 45	Pleiades	Tau	OC	03h 47m	+24° 07'	1.5	120'
M 33	Triangulum Galaxy	Tri	Gx	01h 34m	+30° 40'	6.4	62'
Alp UMi	Polaris	UMi	MS	02h 51m	+89° 20'	2.0	18"

Chart 2

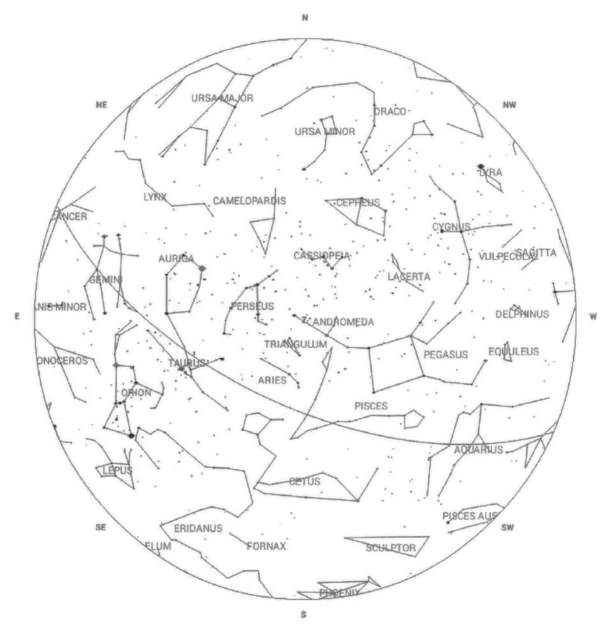

Designation	Name	Con.	Type	R.A.	Dec.	Mag	Size/Sep
Gam And	Almach	And	MS	02h 04m	+42° 20'	2.1	10"
M 31	Andromeda Galaxy	And	Gx	00h 43m	+41° 16'	4.3	156'
NGC 752	Golf Ball Cluster	And	OC	01h 58m	+37° 47'	6.6	75'
NGC 7662	Blue Snowball Nebula	And	PN	23h 26m	+42° 32'	8.6	17"
Pi And		And	MS	00h 38m	+33° 49'	4.4	36"
Lam Ari		Ari	MS	01h 59m	+23° 41'	4.8	37"
Gam Ari	Mesarthim	Ari	MS	01h 54m	+19° 22'	4.6	8"
30 Ari		Ari	MS	02h 37m	+24° 39'	6.5	39"

Designation	Name	Con.	Type	R.A.	Dec.	Mag	Size/Sep
U Cam		Cam	Var/CS	03h 42m	+62° 39'	7.0-7.5	N/A
ST Cam		Cam	Var/CS	04h 51m	+68° 10'	7.0-8.4	N/A
Kemble 1	Kemble's Cascade	Cam	Ast	03h 57m	+63° 04'	5.0	180'
NGC 1502	Jolly Roger Cluster	Cam	OC	04h 08m	+62° 20'	4.1	8'
NGC 457	Owl Cluster	Cas	OC	01h 20m	+58° 17'	5.1	20'
Iot Cas		Cas	MS	02h 29m	+67° 24'	4.5	7"
Sig Cas		Cas	MS	23h 59m	+55° 45'	4.9	3"
Struve 163		Cas	MS	01h 51m	+64° 51'	6.5	35"
Struve 3053		Cas	MS	00h 03m	+66° 06'	5.9	15"
Eta Cas	Achird	Cas	MS	00h 50m	+57° 54'	3.6	13"
M 103		Cas	OC	01h 33m	+60° 39'	6.9	5'
M 52		Cas	OC	23h 25m	+61° 36'	8.2	15'
NGC 281		Cas	OC	00h 53m	+56° 38'	7.4	4'
NGC 559		Cas	OC	01h 30m	+63° 18'	7.4	6'
NGC 659	Ying Yang Cluster	Cas	OC	01h 44m	+60° 40'	7.2	5'
NGC 663		Cas	OC	01h 46m	+61° 14'	6.4	14'
NGC 7789	Herschel's Spiral Cluster	Cas	OC	23h 57m	+56° 43'	7.5	25'
Omi Cep		Cep	MS	23h 19m	+68° 07'	4.8	3"
Gam Cet	Kaffaljidhma	Cet	MS	02h 43m	+03° 14'	3.5	3"
Omi Cet	Mira	Cet	Var	02h 19m	-02° 59'	2.0-10.1	N/A
32 Eri		Eri	MS	03h 54m	-02° 57'	4.4	7"
40 Eri	Keid	Eri	MS	04h 15m	-07° 39'	4.4	83"
NGC 1528	m & m Double Cluster	Per	OC	04h 15m	+51° 13'	6.4	16'
Bet Per	Algol	Per	Var	03h 08m	+40° 57'	2.1-3.4	N/A
Eps Per		Per	MS	03h 58m	+40° 01'	2.9	9"
NGC 1245		Per	OC	03h 15m	+47° 15'	7.7	10'
M 34		Per	OC	02h 42m	+42° 46'	5.8	35'
NGC 869/884	Double Cluster	Per	OC	02h 21m	+57° 08'	4.4	18'
Eta Per		Per	MS	02h 51m	+55° 54'	3.8	29"
Melotte 20	Alpha Persei Moving Cluster	Per	OC	03h 24m	+49° 52'	2.3	300'
Alp Psc	Alrisha	Psc	MS	02h 02m	+02° 46'	3.8	2"
Psi1 Psc		Psc	MS	01h 06m	+21° 28'	5.3	30"
Zet Psc		Psc	MS	01h 14m	+07° 35'	5.2	23"
55 Psc		Psc	MS	00h 40m	+21° 26'	5.4	6"
65 Psc		Psc	MS	00h 50m	+27° 43'	7.0	4"
TX Psc		Psc	Var/CS	23h 46m	+03° 29'	4.5-5.3	N/A
M 45	Pleiades	Tau	OC	03h 47m	+24° 07'	1.5	120'
Melotte 25	Hyades	Tau	OC	04h 27m	+15° 52'	0.8	330'
NGC 1647	Pirate Moon Cluster	Tau	OC	04h 46m	+19° 07'	6.2	40'
M 33	Triangulum Galaxy	Tri	Gx	01h 34m	+30° 40'	6.4	62'
Alp UMi	Polaris	UMi	MS	02h 51m	+89° 20'	2.0	18"

Chart 3

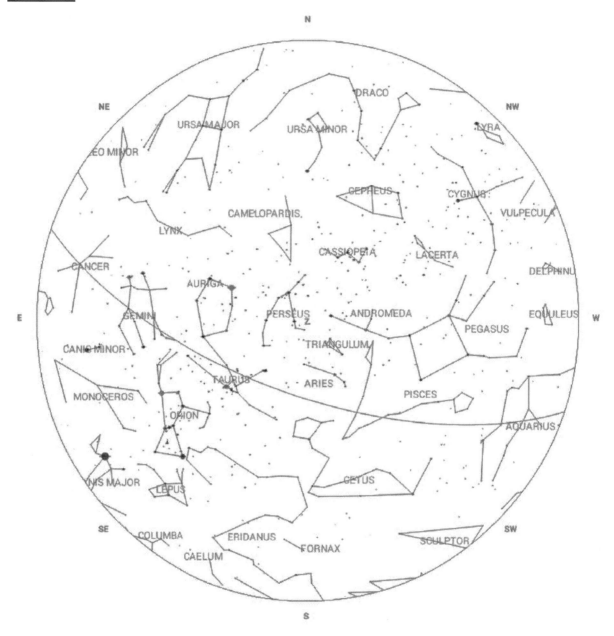

Designation	Name	Con.	Type	R.A.	Dec.	Mag	Size/Sep
Gam And	Almach	And	MS	02h 04m	+42° 20'	2.1	10"
M 31	Andromeda Galaxy	And	Gx	00h 43m	+41° 16'	4.3	156'
Pi And		And	MS	00h 38m	+33° 49'	4.4	36"
Lam Ari		Ari	MS	01h 59m	+23° 41'	4.8	37"
Gam Ari	Mesarthim	Ari	MS	01h 54m	+19° 22'	4.6	8"
30 Ari		Ari	MS	02h 37m	+24° 39'	6.5	39"
M 37		Aur	OC	05h 52m	+32° 33'	6.2	14'
M 36		Aur	OC	05h 36m	+34° 08'	6.5	10'

Designation	Name	Con.	Type	R.A.	Dec.	Mag	Size/Sep
14 Aur		Aur	MS	05h 15m	+32° 41'	5.0	15"
Kemble 1	Kemble's Cascade	Cam	Ast	03h 57m	+63° 04'	5.0	180'
NGC 1502	Jolly Roger Cluster	Cam	OC	04h 08m	+62° 20'	4.1	8'
NGC 457	Owl Cluster	Cas	OC	01h 20m	+58° 17'	5.1	20'
Iot Cas		Cas	MS	02h 29m	+67° 24'	4.5	7"
Struve 3053		Cas	MS	00h 03m	+66° 06'	5.9	15"
Eta Cas	Achird	Cas	MS	00h 50m	+57° 54'	3.6	13"
NGC 663		Cas	OC	01h 46m	+61° 14'	6.4	14'
Gam Cet	Kaffajidhma	Cet	MS	02h 43m	+03° 14'	3.5	3"
Omi Cet	Mira	Cet	Var	02h 19m	-02° 59'	2.0-10.1	N/A
32 Eri		Eri	MS	03h 54m	-02° 57'	4.4	7"
40 Eri	Keid	Eri	MS	04h 15m	-07° 39'	4.4	83"
Sig Ori		Ori	MS	05h 40m	-02° 36'	3.8	42"
Bet Ori	Rigel	Ori	MS	05h 14m	-08° 12'	0.2	10"
Alp Ori	Betelgeuse	Ori	Var	05h 55m	+07° 24'	0.4-1.3	N/A
Eta Ori		Ori	MS	05h 25m	-02° 24'	3.3	2"
Zet Ori	Alnitak	Ori	MS	05h 41m	-01° 57'	1.8	3"
23 Ori		Ori	MS	05h 23m	+03° 33'	5.0	32"
Iot Ori	Nair al Saif	Ori	MS	05h 35m	-05° 55'	2.8	11"
Collinder 70	Epsilon Orionis Cluster	Ori	OC	05h 36m	-01° 00'	0.4	150'
Struve 747		Ori	MS	05h 35m	-05° 55'	4.8	36"
Collinder 72		Ori	OC	05h 35m	-05° 55'	2.5	20'
Del Ori	Mintaka	Ori	MS	05h 33m	-00° 18'	2.1	53"
Lam Ori	Meissa	Ori	MC	05h 36m	+09° 56'	3.4	4"
M 42	Orion Nebula	Ori	Neb	05h 35m	-05° 23'	4.0	40'
NGC 1981	Coal Car Cluster	Ori	OC	05h 35m	-04° 26'	4.2	28'
Collinder 69	Lambda Orionis Cluster	Ori	OC	05h 35m	+09° 56'	2.8	70'
NGC 1528	m & m Double Cluster	Per	OC	04h 15m	+51° 13'	6.4	16'
Bet Per	Algol	Per	Var	03h 08m	+40° 57'	2.1-3.4	N/A
Eps Per		Per	MS	03h 58m	+40° 01'	2.9	9"
M 34		Per	OC	02h 42m	+42° 46'	5.8	35'
NGC 869/884	Double Cluster	Per	OC	02h 21m	+57° 08'	4.4	18'
Eta Per		Per	MS	02h 51m	+55° 54'	3.8	29"
Melotte 20	Alpha Persei Moving Cluster	Per	OC	03h 24m	+49° 52'	2.3	300'
Alp Psc	Alrisha	Psc	MS	02h 02m	+02° 46'	3.8	2"
Psi1 Psc		Psc	MS	01h 06m	+21° 28'	5.3	30"
Zet Psc		Psc	MS	01h 14m	+07° 35'	5.2	23"
55 Psc		Psc	MS	00h 40m	+21° 26'	5.4	6"
M 45	Pleiades	Tau	OC	03h 47m	+24° 07'	1.5	120'
Melotte 25	Hyades	Tau	OC	04h 27m	+15° 52'	0.8	330'
NGC 1647	Pirate Moon Cluster	Tau	OC	04h 46m	+19° 07'	6.2	40'
118 Tau		Tau	MS	05h 29m	+25° 09'	5.5	5"
M 33	Triangulum Galaxy	Tri	Gx	01h 34m	+30° 40'	6.4	62'
Alp UMi	Polaris	UMi	MS	02h 51m	+89° 20'	2.0	18"

Chart 4

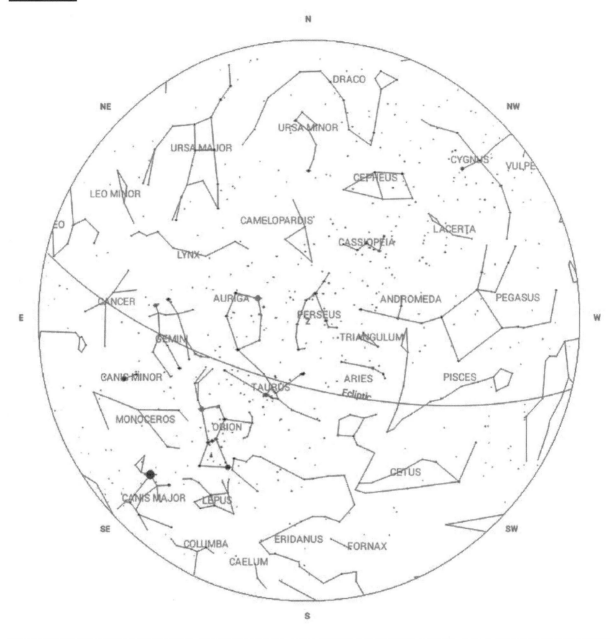

Designation	Name	Con.	Type	R.A.	Dec.	Mag	Size/Sep
Gam And	Almach	And	MS	02h 04m	+42° 20'	2.1	10"
Lam Ari		Ari	MS	01h 59m	+23° 41'	4.8	37"
Gam Ari	Mesarthim	Ari	MS	01h 54m	+19° 22'	4.6	8"
UU Aur		Aur	Var/CS	06h 36m	+38° 27'	5.3-6.5	N/A
The Aur		Aur	MS	06h 00m	+37° 13'	2.7	4"
M 37		Aur	OC	05h 52m	+32° 33'	6.2	14'
14 Aur		Aur	MS	05h 15m	+32° 41'	5.0	15"
Kemble 1	Kemble's Cascade	Cam	Ast	03h 57m	+63° 04'	5.0	180'

Designation	Name	Con.	Type	R.A.	Dec.	Mag	Size/Sep
NGC 1502	Jolly Roger Cluster	Cam	OC	04h 08m	+62° 20'	4.1	8'
NGC 457	Owl Cluster	Cas	OC	01h 20m	+58° 17'	5.1	20'
Iot Cas		Cas	MS	02h 29m	+67° 24'	4.5	7"
NGC 663		Cas	OC	01h 46m	+61° 14'	6.4	14'
Gam Cet	Kaffajidhma	Cet	MS	02h 43m	+03° 14'	3.5	3"
Omi Cet	Mira	Cet	Var	02h 19m	-02° 59'	2.0-10.1	N/A
32 Eri		Eri	MS	03h 54m	-02° 57'	4.4	7"
40 Eri	Keid	Eri	MS	04h 15m	-07° 39'	4.4	83"
M 35		Gem	OC	06h 09m	+24° 21'	5.6	25'
38 Gem		Gem	MS	06h 55m	+13° 11'	4.7	7"
Gam Lep		Lep	MS	05h 45m	-22° 27'	3.6	98"
R Lep	Hind's Crimson Star	Lep	Var/CS	05h 00m	-14° 48'	5.5-11.7	N/A
12 Lyn		Lyn	MS	06h 46m	+59° 27'	4.9	9"
Sig Ori		Ori	MS	05h 40m	-02° 36'	3.8	42"
Bet Ori	Rigel	Ori	MS	05h 14m	-08° 12'	0.2	10"
Alp Ori	Betelgeuse	Ori	Var	05h 55m	+07° 24'	0.4-1.3	N/A
Eta Ori		Ori	MS	05h 25m	-02° 24'	3.3	2"
Zet Ori	Alnitak	Ori	MS	05h 41m	-01° 57'	1.8	3"
23 Ori		Ori	MS	05h 23m	+03° 33'	5.0	32"
Iot Ori	Nair al Saif	Ori	MS	05h 35m	-05° 55'	2.8	11"
Collinder 70	Epsilon Orionis Cluster	Ori	OC	05h 36m	-01° 00'	0.4	150'
Struve 747		Ori	MS	05h 35m	-05° 55'	4.8	36"
Collinder 72		Ori	OC	05h 35m	-05° 55'	2.5	20'
Del Ori	Mintaka	Ori	MS	05h 33m	-00° 18'	2.1	53"
Lam Ori	Meissa	Ori	MC	05h 36m	+09° 56'	3.4	4"
M 42	Orion Nebula	Ori	Neb	05h 35m	-05° 23'	4.0	40'
NGC 1981	Coal Car Cluster	Ori	OC	05h 35m	-04° 26'	4.2	28'
Collinder 69	Lambda Orionis Cluster	Ori	OC	05h 35m	+09° 56'	2.8	70'
NGC 1528	m & m Double Cluster	Per	OC	04h 15m	+51° 13'	6.4	16'
Bet Per	Algol	Per	Var	03h 08m	+40° 57'	2.1-3.4	N/A
Eps Per		Per	MS	03h 58m	+40° 01'	2.9	9"
M 34		Per	OC	02h 42m	+42° 46'	5.8	35'
NGC 869/884	Double Cluster	Per	OC	02h 21m	+57° 08'	4.4	18'
Eta Per		Per	MS	02h 51m	+55° 54'	3.8	29"
Melotte 20	Alpha Persei Moving Cluster	Per	OC	03h 24m	+49° 52'	2.3	300'
Alp Psc	Alrisha	Psc	MS	02h 02m	+02° 46'	3.8	2"
Psi1 Psc		Psc	MS	01h 06m	+21° 28'	5.3	30"
Zet Psc		Psc	MS	01h 14m	+07° 35'	5.2	23"
M 45	Pleiades	Tau	OC	03h 47m	+24° 07'	1.5	120'
Melotte 25	Hyades	Tau	OC	04h 27m	+15° 52'	0.8	330'
NGC 1647	Pirate Moon Cluster	Tau	OC	04h 46m	+19° 07'	6.2	40'
118 Tau		Tau	MS	05h 29m	+25° 09'	5.5	5"
M 33	Triangulum Galaxy	Tri	Gx	01h 34m	+30° 40'	6.4	62'
Alp UMi	Polaris	UMi	MS	02h 51m	+89° 20'	2.0	18"

Chart 5

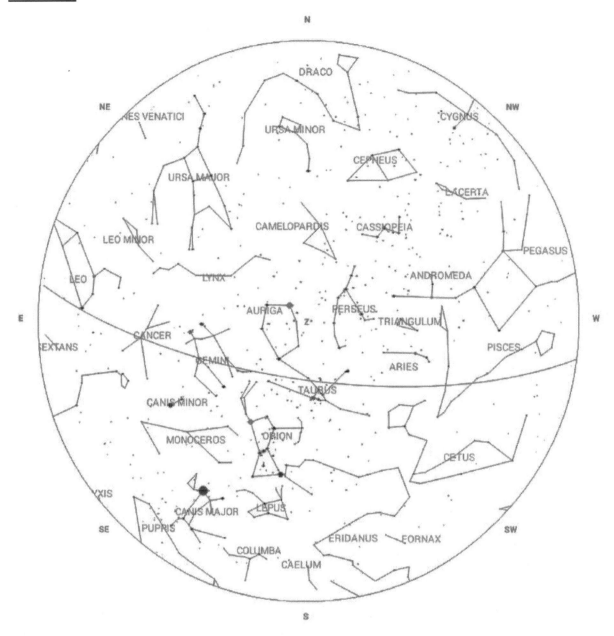

Designation	Name	Con.	Type	R.A.	Dec.	Mag	Size/Sep
Gam And	Almach	And	MS	02h 04m	+42° 20'	2.1	10"
The Aur		Aur	MS	06h 00m	+37° 13'	2.7	4"
14 Aur		Aur	MS	05h 15m	+32° 41'	5.0	15"
U Cam		Cam	Var/CS	03h 42m	+62° 39'	7.0-7.5	N/A
ST Cam		Cam	Var/CS	04h 51m	+68° 10'	7.0-8.4	N/A
Kemble 1	Kemble's Cascade	Cam	Ast	03h 57m	+63° 04'	5.0	180'
NGC 1502	Jolly Roger Cluster	Cam	OC	04h 08m	+62° 20'	4.1	8'
Iot Cas		Cas	MS	02h 29m	+67° 24'	4.5	7"

Designation	Name	Con.	Type	R.A.	Dec.	Mag	Size/Sep
32 Eri		Eri	MS	03h 54m	-02° 57'	4.4	7"
40 Eri	Keid	Eri	MS	04h 15m	-07° 39'	4.4	83"
M 35		Gem	OC	06h 09m	+24° 21'	5.6	25'
38 Gem		Gem	MS	06h 55m	+13° 11'	4.7	7"
Alp Gem	Castor	Gem	MS	07h 35m	+31° 53'	1.6	3"
Del Gem	Wasat	Gem	MS	07h 20m	+21° 59'	3.5	6"
Kap Gem		Gem	MS	07h 44m	+24° 24'	3.6	7"
Gam Lep		Lep	MS	05h 45m	-22° 27'	3.6	98"
R Lep	Hind's Crimson Star	Lep	Var/CS	05h 00m	-14° 48'	5.5-11.7	N/A
12 Lyn		Lyn	MS	06h 46m	+59° 27'	4.9	9"
19 Lyn		Lyn	MS	07h 23m	+55° 17'	5.8	215"
NGC 2353		Mon	OC	07h 15m	-10° 16'	5.2	18'
Bet Mon		Mon	MS	06h 29m	-07° 02'	3.8	7"
NGC 2237	Rosette Nebula	Mon	Neb	06h 32m	+04° 59'	5.5	70'
NGC 2244		Mon	OC	06h 32m	+04° 57'	5.2	29'
NGC 2264	Christmas Tree Cluster	Mon	OC	06h 41m	+09° 54'	4.1	39'
Eps Mon		Mon	MS	06h 24m	+04° 36'	4.3	13"
Sig Ori		Ori	MS	05h 40m	-02° 36'	3.8	42"
Bet Ori	Rigel	Ori	MS	05h 14m	-08° 12'	0.2	10"
Alp Ori	Betelgeuse	Ori	Var	05h 55m	+07° 24'	0.4-1.3	N/A
Eta Ori		Ori	MS	05h 25m	-02° 24'	3.3	2"
Zet Ori	Alnitak	Ori	MS	05h 41m	-01° 57'	1.8	3"
23 Ori		Ori	MS	05h 23m	+03° 33'	5.0	32"
Iot Ori	Nair al Saif	Ori	MS	05h 35m	-05° 55'	2.8	11"
W Ori		Ori	Var/CS	05h 05m	+01° 11'	6.2-7.0	N/A
Collinder 70	Epsilon Orionis Cluster	Ori	OC	05h 36m	-01° 00'	0.4	150'
Struve 747		Ori	MS	05h 35m	-05° 55'	4.8	36"
Collinder 72		Ori	OC	05h 35m	-05° 55'	2.5	20'
Del Ori	Mintaka	Ori	MS	05h 33m	-00° 18'	2.1	53"
Lam Ori	Meissa	Ori	MC	05h 36m	+09° 56'	3.4	4"
M 42	Orion Nebula	Ori	Neb	05h 35m	-05° 23'	4.0	40'
NGC 1981	Coal Car Cluster	Ori	OC	05h 35m	-04° 26'	4.2	28'
Collinder 69	Lambda Orionis Cluster	Ori	OC	05h 35m	+09° 56'	2.8	70'
BL Ori		Ori	Var/CS	06h 26m	+14° 43'	6.3-7.0	N/A
Bet Per	Algol	Per	Var	03h 08m	+40° 57'	2.1-3.4	N/A
Eps Per		Per	MS	03h 58m	+40° 01'	2.9	9"
M 34		Per	OC	02h 42m	+42° 46'	5.8	35'
NGC 869/884	Double Cluster	Per	OC	02h 21m	+57° 08'	4.4	18'
Eta Per		Per	MS	02h 51m	+55° 54'	3.8	29"
Melotte 20	Alpha Persei Moving Cluster	Per	OC	03h 24m	+49° 52'	2.3	300'
M 45	Pleiades	Tau	OC	03h 47m	+24° 07'	1.5	120'
Melotte 25	Hyades	Tau	OC	04h 27m	+15° 52'	0.8	330'
118 Tau		Tau	MS	05h 29m	+25° 09'	5.5	5"
Alp UMi	Polaris	UMi	MS	02h 51m	+89° 20'	2.0	18"

Chart 6

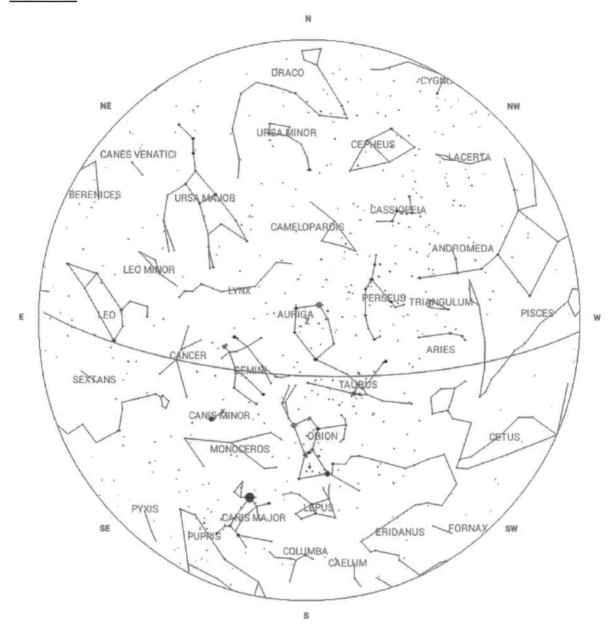

Designation	Name	Con.	Type	R.A.	Dec.	Mag	Size/Sep
UU Aur		Aur	Var/CS	06h 36m	+38° 27'	5.3-6.5	N/A
The Aur		Aur	MS	06h 00m	+37° 13'	2.7	4"
14 Aur		Aur	MS	05h 15m	+32° 41'	5.0	15"
Kemble 1	Kemble's Cascade	Cam	Ast	03h 57m	+63° 04'	5.0	180'
NGC 1502	Jolly Roger Cluster	Cam	OC	04h 08m	+62° 20'	4.1	8'
M 41		CMa	OC	05h 46m	-20° 45'	5.0	39'
Eps CMa	Adhara	CMa	MS	06h 59m	-28° 58'	1.5	8"
NGC 2362	Tau Canis Majoris Cluster	CMa	OC	07h 19m	-24° 57'	3.8	5'

Designation	Name	Con.	Type	R.A.	Dec.	Mag	Size/Sep
Iot Cnc		Cnc	MS	08h 47m	+28° 46'	4.0	30"
M 44	Praesepe	Cnc	OC	08h 40m	+19° 40'	3.9	70'
Zet Cnc	Tegmen	Cnc	MS	08h 12m	+17° 39'	4.7	6"
Phi2 Cnc		Cnc	MS	08h 27m	+26° 56'	5.6	5"
57 Cnc		Cnc	MS	08h 54m	+30° 35'	5.4	56"
X Cnc		Cnc	Var/CS	08h 55m	+17° 14'	5.6-7.5	N/A
32 Eri		Eri	MS	03h 54m	-02° 57'	4.4	7"
40 Eri	Keid	Eri	MS	04h 15m	-07° 39'	4.4	83"
M 35		Gem	OC	06h 09m	+24° 21'	5.6	25'
38 Gem		Gem	MS	06h 55m	+13° 11'	4.7	7"
Alp Gem	Castor	Gem	MS	07h 35m	+31° 53'	1.6	3"
Del Gem	Wasat	Gem	MS	07h 20m	+21° 59'	3.5	6"
Kap Gem		Gem	MS	07h 44m	+24° 24'	3.6	7"
Gam Lep		Lep	MS	05h 45m	-22° 27'	3.6	98"
R Lep	Hind's Crimson Star	Lep	Var/CS	05h 00m	-14° 48'	5.5-11.7	N/A
12 Lyn		Lyn	MS	06h 46m	+59° 27'	4.9	9"
19 Lyn		Lyn	MS	07h 23m	+55° 17'	5.8	215"
NGC 2353		Mon	OC	07h 15m	-10° 16'	5.2	18'
Bet Mon		Mon	MS	06h 29m	-07° 02'	3.8	7"
NGC 2237	Rosette Nebula	Mon	Neb	06h 32m	+04° 59'	5.5	70'
NGC 2244		Mon	OC	06h 32m	+04° 57'	5.2	29'
NGC 2264	Christmas Tree Cluster	Mon	OC	06h 41m	+09° 54'	4.1	39'
Eps Mon		Mon	MS	06h 24m	+04° 36'	4.3	13"
Sig Ori		Ori	MS	05h 40m	-02° 36'	3.8	42"
Bet Ori	Rigel	Ori	MS	05h 14m	-08° 12'	0.2	10"
Alp Ori	Betelgeuse	Ori	Var	05h 55m	+07° 24'	0.4-1.3	N/A
Eta Ori		Ori	MS	05h 25m	-02° 24'	3.3	2"
Zet Ori	Alnitak	Ori	MS	05h 41m	-01° 57'	1.8	3"
23 Ori		Ori	MS	05h 23m	+03° 33'	5.0	32"
Iot Ori	Nair al Saif	Ori	MS	05h 35m	-05° 55'	2.8	11"
Collinder 70	Epsilon Orionis Cluster	Ori	OC	05h 36m	-01° 00'	0.4	150'
Struve 747		Ori	MS	05h 35m	-05° 55'	4.8	36"
Collinder 72		Ori	OC	05h 35m	-05° 55'	2.5	20'
Del Ori	Mintaka	Ori	MS	05h 33m	-00° 18'	2.1	53"
Lam Ori	Meissa	Ori	MC	05h 36m	+09° 56'	3.4	4"
M 42	Orion Nebula	Ori	Neb	05h 35m	-05° 23'	4.0	40'
NGC 1981	Coal Car Cluster	Ori	OC	05h 35m	-04° 26'	4.2	28'
Collinder 69	Lambda Orionis Cluster	Ori	OC	05h 35m	+09° 56'	2.8	70'
Bet Per	Algol	Per	Var	03h 08m	+40° 57'	2.1-3.4	N/A
Eps Per		Per	MS	03h 58m	+40° 01'	2.9	9"
Melotte 20	Alpha Persei Moving Cluster	Per	OC	03h 24m	+49° 52'	2.3	300'
M 45	Pleiades	Tau	OC	03h 47m	+24° 07'	1.5	120'
Melotte 25	Hyades	Tau	OC	04h 27m	+15° 52'	0.8	330'
118 Tau		Tau	MS	05h 29m	+25° 09'	5.5	5"

Chart 7

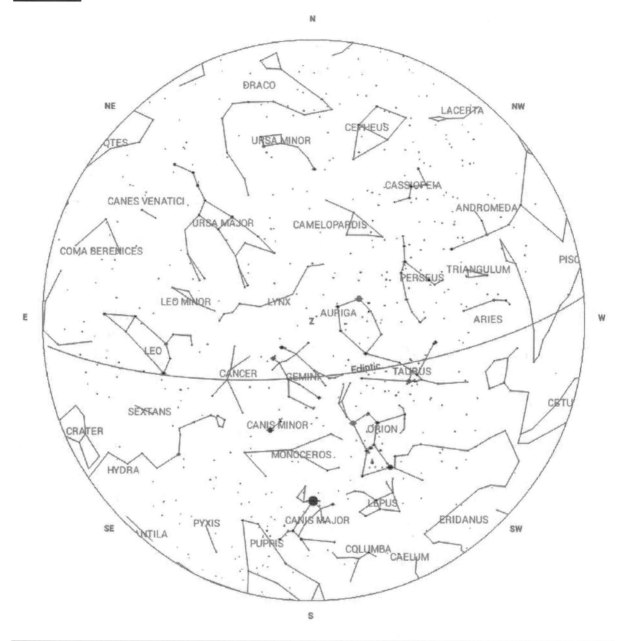

Designation	Name	Con.	Type	R.A.	Dec.	Mag	Size/Sep
14 Aur		Aur	MS	05h 15m	+32° 41'	5.0	15"
The Aur		Aur	MS	06h 00m	+37° 13'	2.7	4"
M 37		Aur	OC	05h 52m	+32° 33'	6.2	14'
UU Aur		Aur	Var/CS	06h 36m	+38° 27'	5.3-6.5	N/A
NGC 1502	Jolly Roger Cluster	Cam	OC	04h 08m	+62° 20'	4.1	8'
Eps CMa	Adhara	CMa	MS	06h 59m	-28° 58'	1.5	8"
M 41		CMa	OC	05h 46m	-20° 45'	5.0	39'
NGC 2362	Tau Canis Majoris Cluster	CMa	OC	07h 19m	-24° 57'	3.8	5'

Designation	Name	Con.	Type	R.A.	Dec.	Mag	Size/Sep
Zet Cnc	Tegmen	Cnc	MS	08h 12m	+17° 39'	4.7	6"
Phi2 Cnc		Cnc	MS	08h 27m	+26° 56'	5.6	5"
Iot Cnc		Cnc	MS	08h 47m	+28° 46'	4.0	30"
57 Cnc		Cnc	MS	08h 54m	+30° 35'	5.4	56"
M 44	Praesepe	Cnc	OC	08h 40m	+19° 40'	3.9	70'
X Cnc		Cnc	Var/CS	08h 55m	+17° 14'	5.6-7.5	N/A
38 Gem		Gem	MS	06h 55m	+13° 11'	4.7	7"
Del Gem	Wasat	Gem	MS	07h 20m	+21° 59'	3.5	6"
Alp Gem	Castor	Gem	MS	07h 35m	+31° 53'	1.6	3"
Kap Gem		Gem	MS	07h 44m	+24° 24'	3.6	7"
M 35		Gem	OC	06h 09m	+24° 21'	5.6	25'
R Leo	Peltier's Variable Star	Leo	Var	09h 48m	+11° 26'	4.4-10.5	N/A
Gam Lep		Lep	MS	05h 45m	-22° 27'	3.6	98"
R Lep	Hind's Crimson Star	Lep	Var/CS	05h 00m	-14° 48'	5.5-11.7	N/A
12 Lyn		Lyn	MS	06h 46m	+59° 27'	4.9	9"
19 Lyn		Lyn	MS	07h 23m	+55° 17'	5.8	215"
38 Lyn		Lyn	MS	09h 19m	+36° 48'	3.8	3"
Eps Mon		Mon	MS	06h 24m	+04° 36'	4.3	13"
Bet Mon		Mon	MS	06h 29m	-07° 02'	3.8	7"
NGC 2237	Rosette Nebula	Mon	Neb	06h 32m	+04° 59'	5.5	70'
NGC 2244		Mon	OC	06h 32m	+04° 57'	5.2	29'
NGC 2264	Christmas Tree Cluster	Mon	OC	06h 41m	+09° 54'	4.1	39'
NGC 2301	Hagrid's Dragon	Mon	OC	06h 52m	+00° 28'	6.3	14'
NGC 2353		Mon	OC	07h 15m	-10° 16'	5.2	18'
Lam Ori	Meissa	Ori	MC	05h 36m	+09° 56'	3.4	4"
Bet Ori	Rigel	Ori	MS	05h 14m	-08° 12'	0.2	10"
23 Ori		Ori	MS	05h 23m	+03° 33'	5.0	32"
Eta Ori		Ori	MS	05h 25m	-02° 24'	3.3	2"
Del Ori	Mintaka	Ori	MS	05h 33m	-00° 18'	2.1	53"
Iot Ori	Nair al Saif	Ori	MS	05h 35m	-05° 55'	2.8	11"
Struve 747		Ori	MS	05h 35m	-05° 55'	4.8	36"
Sig Ori		Ori	MS	05h 40m	-02° 36'	3.8	42"
Zet Ori	Alnitak	Ori	MS	05h 41m	-01° 57'	1.8	3"
M 42	Orion Nebula	Ori	Neb	05h 35m	-05° 23'	4.0	40'
Collinder 72		Ori	OC	05h 35m	-05° 55'	2.5	20'
NGC 1981	Coal Car Cluster	Ori	OC	05h 35m	-04° 26'	4.2	28'
Collinder 69	Lambda Orionis Cluster	Ori	OC	05h 35m	+09° 56'	2.8	70'
Collinder 70	Epsilon Orionis Cluster	Ori	OC	05h 36m	-01° 00'	0.4	150'
Alp Ori	Betelgeuse	Ori	Var	05h 55m	+07° 24'	0.4-1.3	N/A
M 47		Pup	OC	07h 37m	-14° 29'	4.3	25'
118 Tau		Tau	MS	05h 29m	+25° 09'	5.5	5"
Melotte 25	Hyades	Tau	OC	04h 27m	+15° 52'	0.8	330'
NGC 1647	Pirate Moon Cluster	Tau	OC	04h 46m	+19° 07'	6.2	40'

Chart 8

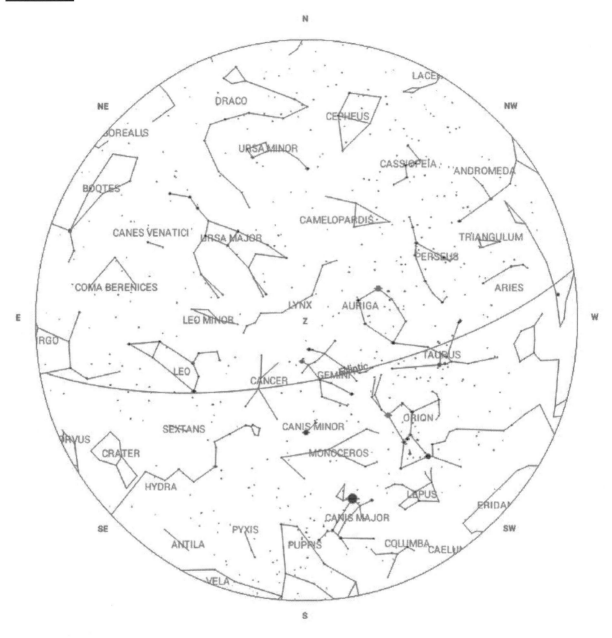

Designation	Name	Con.	Type	R.A.	Dec.	Mag	Size/Sep
14 Aur		Aur	MS	05h 15m	+32° 41'	5.0	15"
The Aur		Aur	MS	06h 00m	+37° 13'	2.7	4"
M 37		Aur	OC	05h 52m	+32° 33'	6.2	14'
UU Aur		Aur	Var/CS	06h 36m	+38° 27'	5.3-6.5	N/A
Eps CMa	Adhara	CMa	MS	06h 59m	-28° 58'	1.5	8"
M 41		CMa	OC	05h 46m	-20° 45'	5.0	39'
NGC 2362	Tau Canis Majoris Cluster	CMa	OC	07h 19m	-24° 57'	3.8	5'
Zet Cnc	Tegmen	Cnc	MS	08h 12m	+17° 39'	4.7	6"

Designation	Name	Con.	Type	R.A.	Dec.	Mag	Size/Sep
Phi2 Cnc		Cnc	MS	08h 27m	+26° 56'	5.6	5"
Iot Cnc		Cnc	MS	08h 47m	+28° 46'	4.0	30"
57 Cnc		Cnc	MS	08h 54m	+30° 35'	5.4	56"
M 44	Praesepe	Cnc	OC	08h 40m	+19° 40'	3.9	70'
X Cnc		Cnc	Var/CS	08h 55m	+17° 14'	5.6-7.5	N/A
38 Gem		Gem	MS	06h 55m	+13° 11'	4.7	7"
Del Gem	Wasat	Gem	MS	07h 20m	+21° 59'	3.5	6"
Alp Gem	Castor	Gem	MS	07h 35m	+31° 53'	1.6	3"
Kap Gem		Gem	MS	07h 44m	+24° 24'	3.6	7"
M 35		Gem	OC	06h 09m	+24° 21'	5.6	25'
Eps Hya		Hya	MS	08h 47m	+06° 25'	3.4	3"
M 48		Hya	OC	08h 14m	-05° 45'	5.5	30'
U Hya		Hya	Var/CS	10h 38m	-13° 23'	4.8-6.5	N/A
R Leo	Peltier's Variable Star	Leo	Var	09h 48m	+11° 26'	4.4-10.5	N/A
Gam Leo	Algieba	Leo	MS	10h 20m	+19° 50'	2.0	5"
54 Leo		Leo	MS	10h 56m	+24° 45'	4.3	6"
12 Lyn		Lyn	MS	06h 46m	+59° 27'	4.9	9"
19 Lyn		Lyn	MS	07h 23m	+55° 17'	5.8	215"
38 Lyn		Lyn	MS	09h 19m	+36° 48'	3.8	3"
Eps Mon		Mon	MS	06h 24m	+04° 36'	4.3	13"
Bet Mon		Mon	MS	06h 29m	-07° 02'	3.8	7"
NGC 2237	Rosette Nebula	Mon	Neb	06h 32m	+04° 59'	5.5	70'
NGC 2244		Mon	OC	06h 32m	+04° 57'	5.2	29'
NGC 2264	Christmas Tree Cluster	Mon	OC	06h 41m	+09° 54'	4.1	39'
NGC 2301	Hagrid's Dragon	Mon	OC	06h 52m	+00° 28'	6.3	14'
NGC 2353		Mon	OC	07h 15m	-10° 16'	5.2	18'
Lam Ori	Meissa	Ori	MC	05h 36m	+09° 56'	3.4	4"
Bet Ori	Rigel	Ori	MS	05h 14m	-08° 12'	0.2	10"
23 Ori		Ori	MS	05h 23m	+03° 33'	5	32"
Eta Ori		Ori	MS	05h 25m	-02° 24'	3.3	2"
Del Ori	Mintaka	Ori	MS	05h 33m	-00° 18'	2.1	53"
Iot Ori	Nair al Saif	Ori	MS	05h 35m	-05° 55'	2.8	11"
Struve 747		Ori	MS	05h 35m	-05° 55'	4.8	36"
Sig Ori		Ori	MS	05h 40m	-02° 36'	3.8	42"
Zet Ori	Alnitak	Ori	MS	05h 41m	-01° 57'	1.8	3"
M 42	Orion Nebula	Ori	Neb	05h 35m	-05° 23'	4.0	40'
Collinder 72		Ori	OC	05h 35m	-05° 55'	2.5	20'
NGC 1981	Coal Car Cluster	Ori	OC	05h 35m	-04° 26'	4.2	28'
Collinder 69	Lambda Orionis Cluster	Ori	OC	05h 35m	+09° 56'	2.8	70'
Collinder 70	Epsilon Orionis Cluster	Ori	OC	05h 36m	-01° 00'	0.4	150'
Alp Ori	Betelgeuse	Ori	Var	05h 55m	+07° 24'	0.4-1.3	N/A
M 47		Pup	OC	07h 37m	-14° 29'	4.3	25'
118 Tau		Tau	MS	05h 29m	+25° 09'	5.5	5"

Chart 9

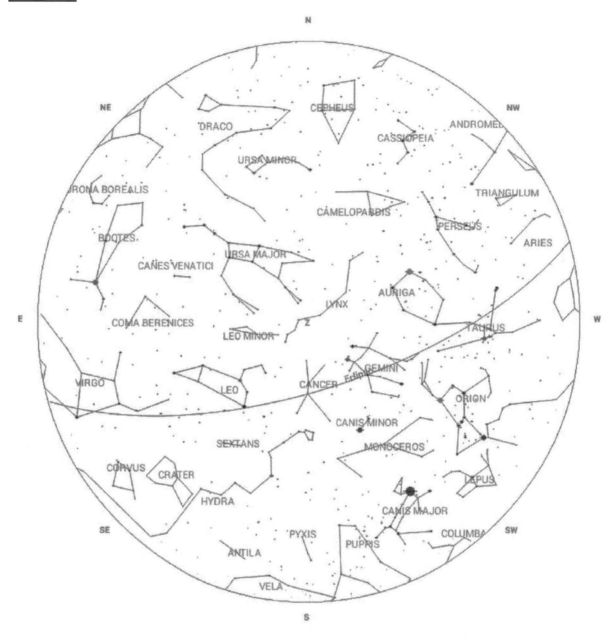

Designation	Name	Con.	Type	R.A.	Dec.	Mag	Size/Sep
The Aur		Aur	MS	06h 00m	+37° 13'	2.7	4"
UU Aur		Aur	Var/CS	06h 36m	+38° 27'	5.3-6.5	N/A
NGC 2403		Cam	Gx	07h 37m	+65° 36'	8.8	20'
Zet Cnc	Tegmen	Cnc	MS	08h 12m	+17° 39'	4.7	6"
Phi2 Cnc		Cnc	MS	08h 27m	+26° 56'	5.6	5"
Iot Cnc		Cnc	MS	08h 47m	+28° 46'	4.0	30"
57 Cnc		Cnc	MS	08h 54m	+30° 35'	5.4	56"
M 44	Praesepe	Cnc	OC	08h 40m	+19° 40'	3.9	70'

Designation	Name	Con.	Type	R.A.	Dec.	Mag	Size/Sep
M 67		Cnc	OC	08h 51m	+11° 48'	7.4	25'
X Cnc		Cnc	Var/CS	08h 55m	+17° 14'	5.6-7.5	N/A
38 Gem		Gem	MS	06h 55m	+13° 11'	4.7	7"
Del Gem	Wasat	Gem	MS	07h 20m	+21° 59'	3.5	6"
Alp Gem	Castor	Gem	MS	07h 35m	+31° 53'	1.6	3"
Kap Gem		Gem	MS	07h 44m	+24° 24'	3.6	7"
M 35		Gem	OC	06h 09m	+24° 21'	5.6	25'
NGC 2392	Eskimo Nebula	Gem	PN	07h 29m	+20° 55'	8.6	47"
NGC 3242	Ghost of Jupiter	Hya	PN	10h 25m	-18° 39'	8.6	40"
Eps Hya		Hya	MS	08h 47m	+06° 25'	3.4	3"
M 48		Hya	OC	08h 14m	-05° 45'	5.5	30'
U Hya		Hya	Var/CS	10h 38m	-13° 23'	4.8-6.5	N/A
R Leo	Peltier's Variable Star	Leo	Var	09h 48m	+11° 26'	4.4-10.5	N/A
Gam Leo	Algieba	Leo	MS	10h 20m	+19° 50'	2.0	5"
54 Leo		Leo	MS	10h 56m	+24° 45'	4.3	6"
M 66		Leo	Gx	11h 20m	+13° 00'	9.7	9'
12 Lyn		Lyn	MS	06h 46m	+59° 27'	4.9	9"
19 Lyn		Lyn	MS	07h 23m	+55° 17'	5.8	215"
38 Lyn		Lyn	MS	09h 19m	+36° 48'	3.8	3"
Eps Mon		Mon	MS	06h 24m	+04° 36'	4.3	13"
NGC 3521		Leo	Gx	11h 06m	-00° 02'	9.9	10'
Bet Mon		Mon	MS	06h 29m	-07° 02'	3.8	7"
Iot Leo		Leo	MS	11h 24m	+10° 32'	3.9	2"
NGC 2237	Rosette Nebula	Mon	Neb	06h 32m	+04° 59'	5.5	70'
NGC 2244		Mon	OC	06h 32m	+04° 57'	5.2	29'
NGC 2264	Christmas Tree Cluster	Mon	OC	06h 41m	+09° 54'	4.1	39'
NGC 2301	Hagrid's Dragon	Mon	OC	06h 52m	+00° 28'	6.3	14'
M 50		Mon	OC	07h 03m	-08° 23'	7.2	14'
NGC 2353		Mon	OC	07h 15m	-10° 16'	5.2	18'
NGC 2506		Mon	OC	08h 00m	-10° 46'	8.9	12'
NGC 2175		Ori	Neb	06h 10m	+20° 29'	6.8	22'
NGC 2169	37 Cluster	Ori	OC	06h 09m	+13° 58'	7.0	5'
NGC 2467		Pup	Neb	07h 52m	-26° 26'	7.1	14'
M 47		Pup	OC	07h 37m	-14° 29'	4.3	25'
M 46		Pup	OC	07h 42m	-14° 48'	6.6	20'
M 93		Pup	OC	07h 45m	-23° 51'	6.5	10'
NGC 2539	Dish Cluster	Pup	OC	08h 11m	-12° 49'	8.0	9'
M 97	Owl Nebula	UMa	PN	11h 15m	+55° 01'	9.7	3'
Xi UMa	Alula Australis	UMa	MS	11h 18m	+31° 32'	4.4	2"
VY UMa		UMa	Var/CS	10h 45m	+67° 25'	5.9-6.5	N/A
M 81	Bode's Galaxy	UMa	Gx	09h 56m	+69° 04'	7.8	22'
M 82	Cigar Galaxy	UMa	Gx	09h 56m	+69° 41'	9.0	9'

Chart 10

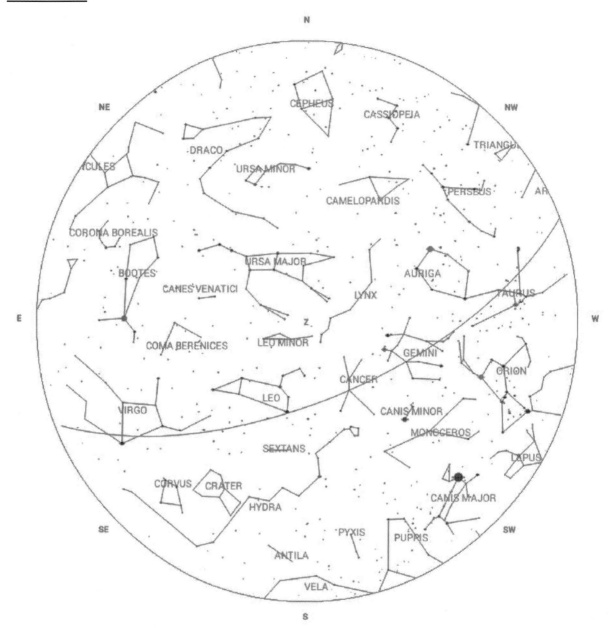

Designation	Name	Con.	Type	R.A.	Dec.	Mag	Size/Sep
NGC 2403		Cam	Gx	07h 37m	+65° 36'	8.8	20'
Zet Cnc	Tegmen	Cnc	MS	08h 12m	+17° 39'	4.7	6"
Phi2 Cnc		Cnc	MS	08h 27m	+26° 56'	5.6	5"
Iot Cnc		Cnc	MS	08h 47m	+28° 46'	4.0	30"
57 Cnc		Cnc	MS	08h 54m	+30° 35'	5.4	56"
M 44	Praesepe	Cnc	OC	08h 40m	+19° 40'	3.9	70'
M 67		Cnc	OC	08h 51m	+11° 48'	7.4	25'
X Cnc		Cnc	Var/CS	08h 55m	+17° 14'	5.6-7.5	N/A
M 64	Black Eye Galaxy	Com	Gx	12h 57m	+21° 41'	9.3	10'

Designation	Name	Con.	Type	R.A.	Dec.	Mag	Size/Sep
Melotte 111	Coma Star Cluster	Com	OC	12h 25m	+26° 06'	2.9	120'
NGC 4725		Com	Gx	12h 50m	+25° 30'	9.9	10'
24 Com		Com	MS	12h 35m	+18° 23'	5.0	20"
M 106		CVn	Gx	12h 19m	+47° 18'	9.1	17'
Alp CVn	Cor Caroli	CVn	MS	12h 56m	+38° 19'	2.9	19"
M 94		CVn	Gx	12h 51m	+41° 07'	8.7	10'
2 CVn		CVn	MS	12h 16m	+40° 40'	5.7	11"
NGC 4656	Hook Galaxy	CVn	Gx	12h 44m	+32° 10'	9.7	9'
NGC 4449		CVn	Gx	12h 28m	+44° 06'	9.5	5'
Y CVn	La Superba	CVn	CS	12h 45m	+45° 26'	5.2-5.5	N/A
NGC 4490	Cocoon Galaxy	CVn	Gx	12h 31m	+41° 39'	9.8	6'
NGC 4631	Whale Galaxy	CVn	Gx	12h 42m	+32° 33'	9.5	13'
RY Dra		Dra	CS	12h 56m	+66° 00'	6.0-8.0	N/A
Del Gem	Wasat	Gem	MS	07h 20m	+21° 59'	3.5	6"
Alp Gem	Castor	Gem	MS	07h 35m	+31° 53'	1.6	3"
Kap Gem		Gem	MS	07h 44m	+24° 24'	3.6	7"
NGC 2392	Eskimo Nebula	Gem	PN	07h 29m	+20° 55'	8.6	47"
M 68		Hya	GC	12h 39m	-26° 45'	7.3	11'
NGC 3242	Ghost of Jupiter	Hya	PN	10h 25m	-18° 39'	8.6	40"
Eps Hya		Hya	MS	08h 47m	+06° 25'	3.4	3"
M 48		Hya	OC	08h 14m	-05° 45'	5.5	30'
U Hya		Hya	Var/CS	10h 38m	-13° 23'	4.8-6.5	N/A
R Leo	Peltier's Variable Star	Leo	Var	09h 48m	+11° 26'	4.4-10.5	N/A
Gam Leo	Algieba	Leo	MS	10h 20m	+19° 50'	2.0	5"
54 Leo		Leo	MS	10h 56m	+24° 45'	4.3	6"
M 66		Leo	Gx	11h 20m	+13° 00'	9.7	9'
NGC 3521		Leo	Gx	11h 06m	-00° 02'	9.9	10'
Iot Leo		Leo	MS	11h 24m	+10° 32'	3.9	2"
19 Lyn		Lyn	MS	07h 23m	+55° 17'	5.8	215"
38 Lyn		Lyn	MS	09h 19m	+36° 48'	3.8	3"
M 40	Winnecke 4	UMa	MS	12h 22m	+58° 05'	9.6	
M 97	Owl Nebula	UMa	PN	11h 15m	+55° 01'	9.7	3'
Xi UMa	Alula Australis	UMa	MS	11h 18m	+31° 32'	4.4	2"
VY UMa		UMa	Var/CS	10h 45m	+67° 25'	5.9-6.5	N/A
M 81	Bode's Galaxy	UMa	Gx	09h 56m	+69° 04'	7.8	22'
M 82	Cigar Galaxy	UMa	Gx	09h 56m	+69° 41'	9.0	9'
M 87		Vir	Gx	12h 31m	+12° 23'	9.6	8'
M 104	Sombrero Galaxy	Vir	Gx	12h 40m	-11° 37'	9.1	9'
M 49		Vir	Gx	12h 30m	+08° 00'	9.3	9'
M 60		Vir	Gx	12h 44m	+11° 33'	9.8	7'
M 86		Vir	Gx	12h 26m	+12° 57'	9.8	10'
Gam Vir	Porrima	Vir	MS	12h 42m	-01° 27'	2.7	2"
SS Vir		Vir	Var/CS	12h 25m	+00° 48'	6.0-9.6	6.0-9.6

Chart 11

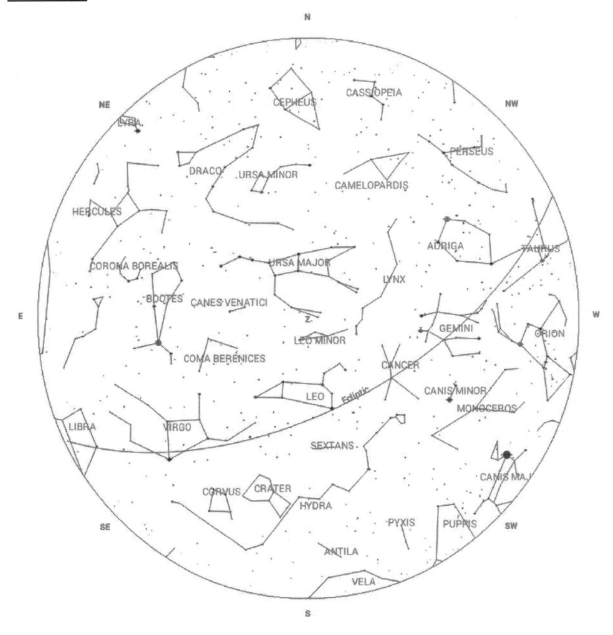

Designation	Name	Con.	Type	R.A.	Dec.	Mag	Size/Sep
Zet Cnc	Tegmen	Cnc	MS	08h 12m	+17° 39'	4.7	6"
Phi2 Cnc		Cnc	MS	08h 27m	+26° 56'	5.6	5"
Iot Cnc		Cnc	MS	08h 47m	+28° 46'	4.0	30"
57 Cnc		Cnc	MS	08h 54m	+30° 35'	5.4	56"
M 44	Praesepe	Cnc	OC	08h 40m	+19° 40'	3.9	70'
M 67		Cnc	OC	08h 51m	+11° 48'	7.4	25'
X Cnc		Cnc	Var/CS	08h 55m	+17° 14'	5.6-7.5	N/A
M 53		Com	GC	13h 13m	+18° 10'	7.7	13'
M 64	Black Eye Galaxy	Com	Gx	12h 57m	+21° 41'	9.3	10'

Designation	Name	Con.	Type	R.A.	Dec.	Mag	Size/Sep
Melotte 111	Coma Star Cluster	Com	OC	12h 25m	+26° 06'	2.9	120'
NGC 4725		Com	Gx	12h 50m	+25° 30'	9.9	10'
24 Com		Com	MS	12h 35m	+18° 23'	5.0	20"
M 51	Whirlpool Galaxy	CVn	Gx	13h 30m	+47° 12'	8.7	10'
M 63	Sunflower Galaxy	CVn	Gx	13h 16m	+42° 02'	9.3	12'
M 106		CVn	Gx	12h 19m	+47° 18'	9.1	17'
M 3		CVn	GC	13h 42m	+28° 23'	6.3	18'
Alp CVn	Cor Caroli	CVn	MS	12h 56m	+38° 19'	2.9	19"
M 94		CVn	Gx	12h 51m	+41° 07'	8.7	10'
2 CVn		CVn	MS	12h 16m	+40° 40'	5.7	11"
NGC 4656	Hook Galaxy	CVn	Gx	12h 44m	+32° 10'	9.7	9'
NGC 4449		CVn	Gx	12h 28m	+44° 06'	9.5	5'
Y CVn	La Superba	CVn	CS	12h 45m	+45° 26'	5.2-5.5	N/A
NGC 4490	Cocoon Galaxy	CVn	Gx	12h 31m	+41° 39'	9.8	6'
NGC 4631	Whale Galaxy	CVn	Gx	12h 42m	+32° 33'	9.5	13'
RY Dra		Dra	CS	12h 56m	+66° 00'	6.0-8.0	N/A
M 83		Hya	Gx	13h 37m	-29° 52'	7.8	14'
M 68		Hya	GC	12h 39m	-26° 45'	7.3	11'
NGC 3242	Ghost of Jupiter	Hya	PN	10h 25m	-18° 39'	8.6	40"
Eps Hya		Hya	MS	08h 47m	+06° 25'	3.4	3"
M 48		Hya	OC	08h 14m	-05° 45'	5.5	30'
U Hya		Hya	Var/CS	10h 38m	-13° 23'	4.8-6.5	N/A
R Leo	Peltier's Variable Star	Leo	Var	09h 48m	+11° 26'	4.4-10.5	N/A
Gam Leo	Algieba	Leo	MS	10h 20m	+19° 50'	2.0	5"
54 Leo		Leo	MS	10h 56m	+24° 45'	4.3	6"
M 66		Leo	Gx	11h 20m	+13° 00'	9.7	9'
NGC 3521		Leo	Gx	11h 06m	-00° 02'	9.9	10'
Iot Leo		Leo	MS	11h 24m	+10° 32'	3.9	2"
38 Lyn		Lyn	MS	09h 19m	+36° 48'	3.8	3"
M 40	Winnecke 4	UMa	MS	12h 22m	+58° 05'	9.6	
M 97	Owl Nebula	UMa	PN	11h 15m	+55° 01'	9.7	3'
Zet UMa	Mizar & Alcor	UMa	MS	13h 24m	+54° 56'	2.1	711"
Xi UMa	Alula Australis	UMa	MS	11h 18m	+31° 32'	4.4	2"
VY UMa		UMa	Var/CS	10h 45m	+67° 25'	5.9-6.5	N/A
M 81	Bode's Galaxy	UMa	Gx	09h 56m	+69° 04'	7.8	22'
M 82	Cigar Galaxy	UMa	Gx	09h 56m	+69° 41'	9.0	9'
M 87		Vir	Gx	12h 31m	+12° 23'	9.6	8'
M 104	Sombrero Galaxy	Vir	Gx	12h 40m	-11° 37'	9.1	9'
M 49		Vir	Gx	12h 30m	+08° 00'	9.3	9'
M 60		Vir	Gx	12h 44m	+11° 33'	9.8	7'
The Vir		Vir	MS	13h 10m	-05° 32'	4.4	70"
M 86		Vir	Gx	12h 26m	+12° 57'	9.8	10'
Gam Vir	Porrima	Vir	MS	12h 42m	-01° 27'	2.7	2"
SS Vir		Vir	Var/CS	12h 25m	+00° 48'	6.0-9.6	6.0-9.6

Chart 12

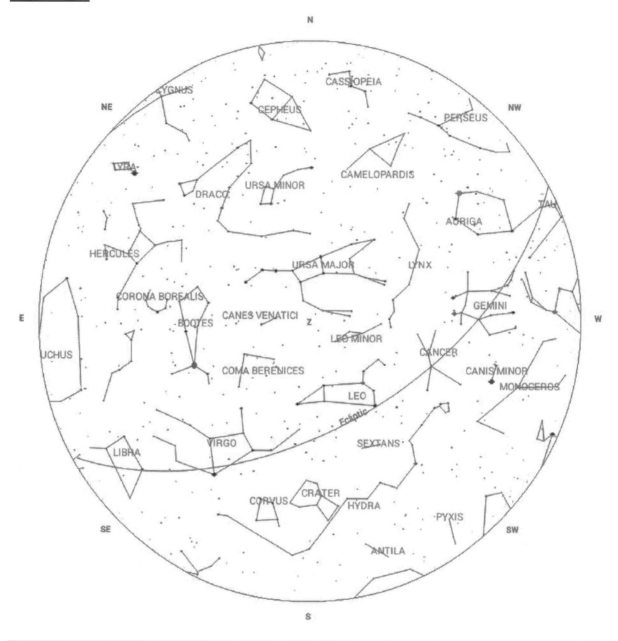

Designation	Name	Con.	Type	R.A.	Dec.	Mag	Size/Sep
Eps Boo	Izar	Boo	MS	14h 45m	+27° 04'	2.4	3"
Xi Boo		Boo	MS	14h 51m	+19° 06'	4.6	6"
Kap Boo	Asellus Tertius	Boo	MS	14h 14m	+51° 47'	4.5	13"
Pi Boo		Boo	MS	14h 41m	+16° 25'	4.5	6"
39 Boo		Boo	MS	14h 50m	+48° 43'	5.7	3"
Struve 1835		Boo	MS	14h 23m	+08° 27'	4.9	6"
M 53		Com	GC	13h 13m	+18° 10'	7.7	13'
M 64	Black Eye Galaxy	Com	Gx	12h 57m	+21° 41'	9.3	10'
Melotte 111	Coma Star Cluster	Com	OC	12h 25m	+26° 06'	2.9	120'

Designation	Name	Con.	Type	R.A.	Dec.	Mag	Size/Sep
24 Com		Com	MS	12h 35m	+18° 23'	5.0	20"
Del Crv	Algorab	Crv	MS	12h 30m	-16° 31'	5.7	24"
Struve 1669		Crv	MS	12h 41m	-13° 01'	5.2	5"
M 51	Whirlpool Galaxy	CVn	Gx	13h 30m	+47° 12'	8.7	10'
M 63	Sunflower Galaxy	CVn	Gx	13h 16m	+42° 02'	9.3	12'
M 106		CVn	Gx	12h 19m	+47° 18'	9.1	17'
M 3		CVn	GC	13h 42m	+28° 23'	6.3	18'
Alp CVn	Cor Caroli	CVn	MS	12h 56m	+38° 19'	2.9	19"
M 94		CVn	Gx	12h 51m	+41° 07'	8.7	10'
2 CVn		CVn	MS	12h 16m	+40° 40'	5.7	11"
NGC 4656	Hook Galaxy	CVn	Gx	12h 44m	+32° 10'	9.7	9'
NGC 4449		CVn	Gx	12h 28m	+44° 06'	9.5	5'
Y CVn	La Superba	CVn	CS	12h 45m	+45° 26'	5.2-5.5	N/A
NGC 4490	Cocoon Galaxy	CVn	Gx	12h 31m	+41° 39'	9.8	6'
NGC 4631	Whale Galaxy	CVn	Gx	12h 42m	+32° 33'	9.5	13'
RY Dra		Dra	CS	12h 56m	+66° 00'	6.0-8.0	N/A
M 83		Hya	Gx	13h 37m	-29° 52'	7.8	14'
54 Hya		Hya	MS	14h 46m	-25° 27'	5.2	9"
M 68		Hya	GC	12h 39m	-26° 45'	7.3	11'
NGC 3242	Ghost of Jupiter	Hya	PN	10h 25m	-18° 39'	8.6	40"
U Hya		Hya	Var/CS	10h 38m	-13° 23'	4.8-6.5	N/A
R Leo	Peltier's Variable Star	Leo	Var	09h 48m	+11° 26'	4.4-10.5	N/A
Gam Leo	Algieba	Leo	MS	10h 20m	+19° 50'	2.0	5"
54 Leo		Leo	MS	10h 56m	+24° 45'	4.3	6"
M 66		Leo	Gx	11h 20m	+13° 00'	9.7	9'
Iot Leo		Leo	MS	11h 24m	+10° 32'	3.9	2"
38 Lyn		Lyn	MS	09h 19m	+36° 48'	3.8	3"
M 101	Pinwheel Galaxy	UMa	Gx	14h 03m	+54° 21'	8.4	22'
M 40	Winnecke 4	UMa	MS	12h 22m	+58° 05'	9.6	
M 97	Owl Nebula	UMa	PN	11h 15m	+55° 01'	9.7	3'
Zet UMa	Mizar & Alcor	UMa	MS	13h 24m	+54° 56'	2.1	711"
Xi UMa	Alula Australis	UMa	MS	11h 18m	+31° 32'	4.4	2"
VY UMa		UMa	Var/CS	10h 45m	+67° 25'	5.9-6.5	N/A
M 81	Bode's Galaxy	UMa	Gx	09h 56m	+69° 04'	7.8	22'
M 82	Cigar Galaxy	UMa	Gx	09h 56m	+69° 41'	9.0	9'
M 87		Vir	Gx	12h 31m	+12° 23'	9.6	8'
M 104	Sombrero Galaxy	Vir	Gx	12h 40m	-11° 37'	9.1	9'
M 49		Vir	Gx	12h 30m	+08° 00'	9.3	9'
M 60		Vir	Gx	12h 44m	+11° 33'	9.8	7'
The Vir		Vir	MS	13h 10m	-05° 32'	4.4	70"
M 86		Vir	Gx	12h 26m	+12° 57'	9.8	10'
Gam Vir	Porrima	Vir	MS	12h 42m	-01° 27'	2.7	2"
SS Vir		Vir	Var/CS	12h 25m	+00° 48'	6.0-9.6	6.0-9.6

Chart 13

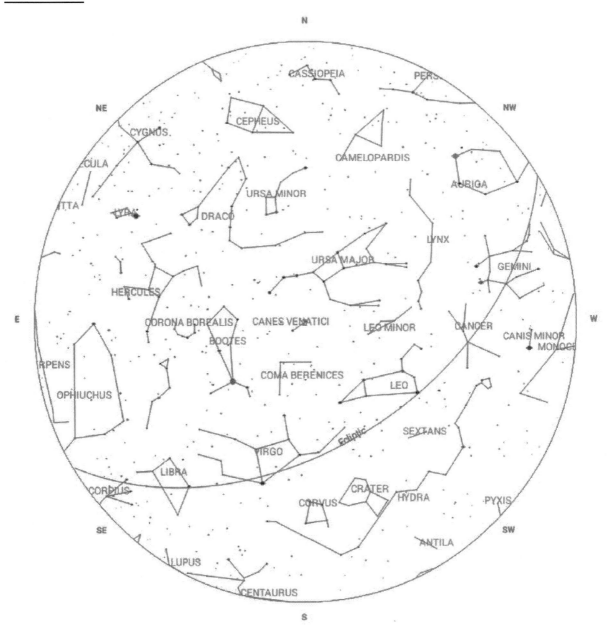

Designation	Name	Con.	Type	R.A.	Dec.	Mag	Size/Sep
Eps Boo	Izar	Boo	MS	14h 45m	+27° 04'	2.4	3"
Xi Boo		Boo	MS	14h 51m	+19° 06'	4.6	6"
Kap Boo	Asellus Tertius	Boo	MS	14h 14m	+51° 47'	4.5	13"
Pi Boo		Boo	MS	14h 41m	+16° 25'	4.5	6"
39 Boo		Boo	MS	14h 50m	+48° 43'	5.7	3"
Struve 1835		Boo	MS	14h 23m	+08° 27'	4.9	6"
Mu Boo	Alkalurops	Boo	MS	15h 24m	+37° 23'	4.3	108"
44 Boo		Boo	MS	15h 04m	+47° 39'	4.8	2"
M 53		Com	GC	13h 13m	+18° 10'	7.7	13'

Designation	Name	Con.	Type	R.A.	Dec.	Mag	Size/Sep
Del Boo		Boo	MS	15h 16m	+33° 19'	3.5	105"
Nu Boo		Boo	MS	15h 31m	+40° 50'	5.0	15'
M 64	Black Eye Galaxy	Com	Gx	12h 57m	+21° 41'	9.3	10'
Melotte 111	Coma Star Cluster	Com	OC	12h 25m	+26° 06'	2.9	120'
24 Com		Com	MS	12h 35m	+18° 23'	5.0	20"
Del Crv	Algorab	Crv	MS	12h 30m	-16° 31'	5.7	24"
Struve 1669		Crv	MS	12h 41m	-13° 01'	5.2	5"
M 51	Whirlpool Galaxy	CVn	Gx	13h 30m	+47° 12'	8.7	10'
M 63	Sunflower Galaxy	CVn	Gx	13h 16m	+42° 02'	9.3	12'
M 106		CVn	Gx	12h 19m	+47° 18'	9.1	17'
M 3		CVn	GC	13h 42m	+28° 23'	6.3	18'
Alp CVn	Cor Caroli	CVn	MS	12h 56m	+38° 19'	2.9	19"
M 94		CVn	Gx	12h 51m	+41° 07'	8.7	10'
2 CVn		CVn	MS	12h 16m	+40° 40'	5.7	11"
NGC 4656	Hook Galaxy	CVn	Gx	12h 44m	+32° 10'	9.7	9'
NGC 4449		CVn	Gx	12h 28m	+44° 06'	9.5	5'
Y CVn	La Superba	CVn	CS	12h 45m	+45° 26'	5.2-5.5	N/A
Zet CrB		CrB	MS	15h 39m	+36° 38'	4.6	6"
R CrB	Fade Out Star	CrB	Var	15h 49m	+28° 09'	5.7-14.8	N/A
NGC 4631	Whale Galaxy	CVn	Gx	12h 42m	+32° 33'	9.5	13'
RY Dra		Dra	CS	12h 56m	+66° 00'	6.0-8.0	N/A
M 83		Hya	Gx	13h 37m	-29° 52'	7.8	14'
54 Hya		Hya	MS	14h 46m	-25° 27'	5.2	9"
M 68		Hya	GC	12h 39m	-26° 45'	7.3	11'
NGC 3242	Ghost of Jupiter	Hya	PN	10h 25m	-18° 39'	8.6	40"
U Hya		Hya	Var/CS	10h 38m	-13° 23'	4.8-6.5	N/A
Gam Leo	Algieba	Leo	MS	10h 20m	+19° 50'	2.0	5"
54 Leo		Leo	MS	10h 56m	+24° 45'	4.3	6"
M 66		Leo	Gx	11h 20m	+13° 00'	9.7	9'
Iot Leo		Leo	MS	11h 24m	+10° 32'	3.9	2"
M 5		Ser	GC	15h 19m	+02° 05'	5.7	23'
Del Ser		Ser	MS	15h 35m	+10° 32'	4.2	4"
M 101	Pinwheel Galaxy	UMa	Gx	14h 03m	+54° 21'	8.4	22'
M 40	Winnecke 4	UMa	MS	12h 22m	+58° 05'	9.6	
M 97	Owl Nebula	UMa	PN	11h 15m	+55° 01'	9.7	3'
Zet UMa	Mizar & Alcor	UMa	MS	13h 24m	+54° 56'	2.1	711"
Xi UMa	Alula Australis	UMa	MS	11h 18m	+31° 32'	4.4	2"
VY UMa		UMa	Var/CS	10h 45m	+67° 25'	5.9-6.5	N/A
M 87		Vir	Gx	12h 31m	+12° 23'	9.6	8'
M 104	Sombrero Galaxy	Vir	Gx	12h 40m	-11° 37'	9.1	9'
M 49		Vir	Gx	12h 30m	+08° 00'	9.3	9'
The Vir		Vir	MS	13h 10m	-05° 32'	4.4	70"
Gam Vir	Porrima	Vir	MS	12h 42m	-01° 27'	2.7	2"
SS Vir		Vir	Var/CS	12h 25m	+00° 48'	6.0-9.6	6.0-9.6

Chart 14

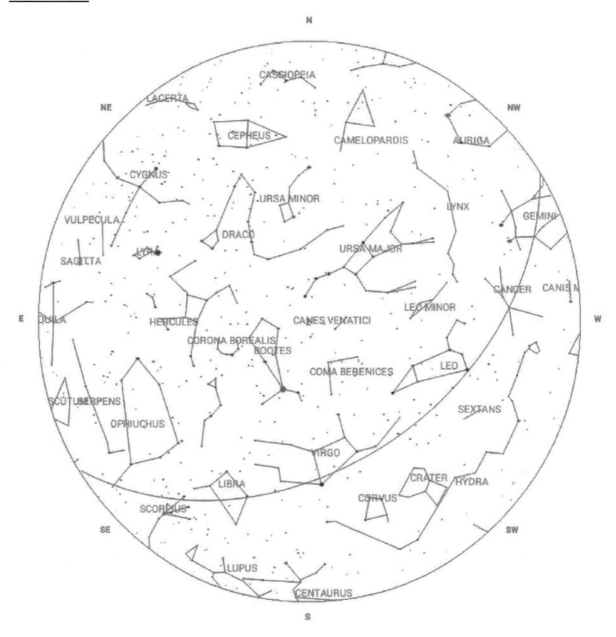

Designation	Name	Con.	Type	R.A.	Dec.	Mag	Size/Sep
Eps Boo	Izar	Boo	MS	14h 45m	+27° 04'	2.4	3"
Xi Boo		Boo	MS	14h 51m	+19° 06'	4.6	6"
Kap Boo	Asellus Tertius	Boo	MS	14h 14m	+51° 47'	4.5	13"
Pi Boo		Boo	MS	14h 41m	+16° 25'	4.5	6"
39 Boo		Boo	MS	14h 50m	+48° 43'	5.7	3"
Struve 1835		Boo	MS	14h 23m	+08° 27'	4.9	6"
Mu Boo	Alkalurops	Boo	MS	15h 24m	+37° 23'	4.3	108"
44 Boo		Boo	MS	15h 04m	+47° 39'	4.8	2"
Del Boo		Boo	MS	15h 16m	+33° 19'	3.5	105"

Designation	Name	Con.	Type	R.A.	Dec.	Mag	Size/Sep
Nu Boo		Boo	MS	15h 31m	+40° 50'	5.0	15'
M 53		Com	GC	13h 13m	+18° 10'	7.7	13'
M 64	Black Eye Galaxy	Com	Gx	12h 57m	+21° 41'	9.3	10'
Melotte 111	Coma Star Cluster	Com	OC	12h 25m	+26° 06'	2.9	120'
24 Com		Com	MS	12h 35m	+18° 23'	5.0	20"
Zet CrB		CrB	MS	15h 39m	+36° 38'	4.6	6"
Sig CrB		CrB	MS	16h 15m	+33° 52'	5.7	7"
R CrB	Fade Out Star	CrB	Var	15h 49m	+28° 09'	5.7-14.8	N/A
T CrB	Blaze Star	CrB	RN	16h 00m	+25° 55'	2.0-10.8	N/A
Del Crv	Algorab	Crv	MS	12h 30m	-16° 31'	5.7	24"
Struve 1669		Crv	MS	12h 41m	-13° 01'	5.2	5"
M 51	Whirlpool Galaxy	CVn	Gx	13h 30m	+47° 12'	8.7	10'
M 63	Sunflower Galaxy	CVn	Gx	13h 16m	+42° 02'	9.3	12'
M 106		CVn	Gx	12h 19m	+47° 18'	9.1	17'
M 3		CVn	GC	13h 42m	+28° 23'	6.3	18'
Alp CVn	Cor Caroli	CVn	MS	12h 56m	+38° 19'	2.9	19"
M 94		CVn	Gx	12h 51m	+41° 07'	8.7	10'
2 CVn		CVn	MS	12h 16m	+40° 40'	5.7	11"
NGC 4449		CVn	Gx	12h 28m	+44° 06'	9.5	5'
Y CVn	La Superba	CVn	CS	12h 45m	+45° 26'	5.2-5.5	N/A
NGC 4631	Whale Galaxy	CVn	Gx	12h 42m	+32° 33'	9.5	13'
RY Dra		Dra	CS	12h 56m	+66° 00'	6.0-8.0	N/A
16/17 Dra		Dra	MS	16h 36m	+52° 55'	5.1	90"
M 13	Keystone Cluster	Her	GC	16h 42m	+36° 27'	5.8	20'
Kap Her	Marfik	Her	MS	16h 41m	+31° 36'	5.0	28"
NGC 6229		Her	GC	16h 47m	+47° 32'	9.4	4'
Iot Leo		Leo	MS	11h 24m	+10° 32'	3.9	2"
Alp Lib	Zuben Elgenubi	Lib	MS	14h 51m	-16° 02'	2.8	230"
NGC 5897	Ghost Globular	Lib	GC	15h 17m	-21° 01'	8.4	11'
Bet Lib	The Emerald Star	Lib	*	15h 14m	-09° 23'	2.6	N/A
Struve 1962		Lib	MS	15h 39m	-08° 47'	5.4	12"
M 5		Ser	GC	15h 19m	+02° 05'	5.7	23'
Del Ser		Ser	MS	15h 35m	+10° 32'	4.2	4"
M 101	Pinwheel Galaxy	UMa	Gx	14h 03m	+54° 21'	8.4	22'
M 40	Winnecke 4	UMa	MS	12h 22m	+58° 05'	9.6	
Zet UMa	Mizar & Alcor	UMa	MS	13h 24m	+54° 56'	2.1	711"
Xi UMa	Alula Australis	UMa	MS	11h 18m	+31° 32'	4.4	2"
M 87		Vir	Gx	12h 31m	+12° 23'	9.6	8'
M 104	Sombrero Galaxy	Vir	Gx	12h 40m	-11° 37'	9.1	9'
M 49		Vir	Gx	12h 30m	+08° 00'	9.3	9'
The Vir		Vir	MS	13h 10m	-05° 32'	4.4	70"
Gam Vir	Porrima	Vir	MS	12h 42m	-01° 27'	2.7	2"
SS Vir		Vir	Var/CS	12h 25m	+00° 48'	6.0-9.6	6.0-9.6

Chart 15

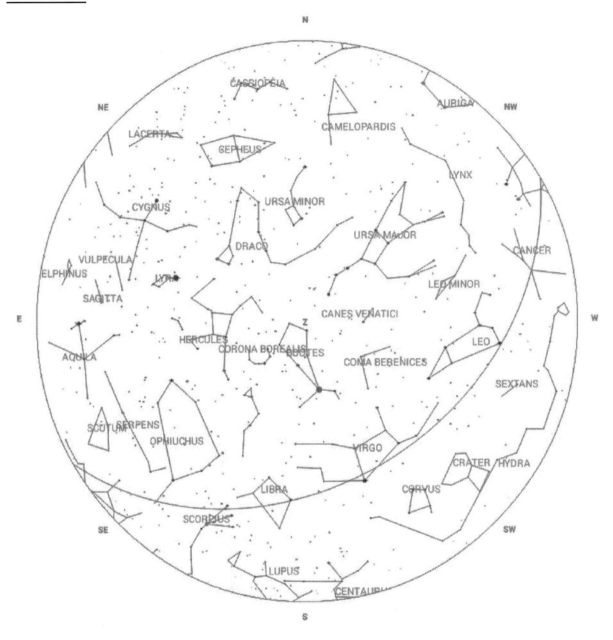

Designation	Name	Con.	Type	R.A.	Dec.	Mag	Size/Sep
Eps Boo	Izar	Boo	MS	14h 45m	+27° 04'	2.4	3"
Xi Boo		Boo	MS	14h 51m	+19° 06'	4.6	6"
Kap Boo	Asellus Tertius	Boo	MS	14h 14m	+51° 47'	4.5	13"
Pi Boo		Boo	MS	14h 41m	+16° 25'	4.5	6"
39 Boo		Boo	MS	14h 50m	+48° 43'	5.7	3"
Struve 1835		Boo	MS	14h 23m	+08° 27'	4.9	6"
Mu Boo	Alkalurops	Boo	MS	15h 24m	+37° 23'	4.3	108"
44 Boo		Boo	MS	15h 04m	+47° 39'	4.8	2"
Del Boo		Boo	MS	15h 16m	+33° 19'	3.5	105"

Designation	Name	Con.	Type	R.A.	Dec.	Mag	Size/Sep
Nu Boo		Boo	MS	15h 31m	+40° 50'	5.0	15'
M 53		Com	GC	13h 13m	+18° 10'	7.7	13'
Melotte 111	Coma Star Cluster	Com	OC	12h 25m	+26° 06'	2.9	120'
24 Com		Com	MS	12h 35m	+18° 23'	5.0	20"
Zet CrB		CrB	MS	15h 39m	+36° 38'	4.6	6"
Sig CrB		CrB	MS	16h 15m	+33° 52'	5.7	7"
R CrB	Fade Out Star	CrB	Var	15h 49m	+28° 09'	5.7-14.8	N/A
T CrB	Blaze Star	CrB	RN	16h 00m	+25° 55'	2.0-10.8	N/A
M 3		CVn	GC	13h 42m	+28° 23'	6.3	18'
Alp CVn	Cor Caroli	CVn	MS	12h 56m	+38° 19'	2.9	19"
2 CVn		CVn	MS	12h 16m	+40° 40'	5.7	11"
Y CVn	La Superba	CVn	CS	12h 45m	+45° 26'	5.2-5.5	N/A
RY Dra		Dra	CS	12h 56m	+66° 00'	6.0-8.0	N/A
16/17 Dra		Dra	MS	16h 36m	+52° 55'	5.1	90"
Nu Dra	Kuma	Dra	MS	17h 32m	+55° 10'	4.9	63"
Mu Dra		Dra	MS	17h 05m	+54° 28'	5.8	2"
Psi Dra		Dra	MS	17h 42m	+72° 09'	4.6	30"
M 13	Keystone Cluster	Her	GC	16h 42m	+36° 27'	5.8	20'
Kap Her	Marfik	Her	MS	16h 41m	+31° 36'	5.0	28"
M 92		Her	GC	17h 17m	+43° 08'	6.5	14'
Alp Her	Rasalgethi	Her	MS	17h 15m	+14° 23'	3.1	5"
Del Her	Sarin	Her	MS	17h 15m	+24° 50'	3.1	14"
Rho Her		Her	MS	17h 24m	+37° 09'	4.2	4"
Alp Lib	Zuben Elgenubi	Lib	MS	14h 51m	-16° 02'	2.8	230"
Bet Lib	The Emerald Star	Lib	*	15h 14m	-09° 23'	2.6	N/A
Struve 1962		Lib	MS	15h 39m	-08° 47'	5.4	12"
IC 4665	Summer Beehive	Oph	OC	17h 46m	+05° 43'	5.3	70'
M 10		Oph	GC	16h 57m	-04° 06'	6.6	20'
M 14		Oph	GC	17h 38m	-03° 15'	7.6	11'
M 12		Oph	GC	16h 47m	-01° 57'	6.1	16'
Rho Oph		Oph	MS	16h 26m	-23° 27'	4.6	3"
M 19		Oph	GC	17h 03m	-26° 16'	6.8	17'
M 62		Oph	GC	17h 01m	-30° 07'	6.4	15'
36 Oph		Oph	MS	17h 15m	-26° 36'	4.3	730"
Omi Oph		Oph	MS	17h 18m	-24° 17'	5.1	10"
61 Oph		Oph	MS	17h 45m	+02° 35'	6.2	21"
M 5		Ser	GC	15h 19m	+02° 05'	5.7	23'
Del Ser		Ser	MS	15h 35m	+10° 32'	4.2	4"
Zet UMa	Mizar & Alcor	UMa	MS	13h 24m	+54° 56'	2.1	711"
The Vir		Vir	MS	13h 10m	-05° 32'	4.4	70"
Gam Vir	Porrima	Vir	MS	12h 42m	-01° 27'	2.7	2"
SS Vir		Vir	Var/CS	12h 25m	+00° 48'	6.0-9.6	6.0-9.6

Chart 16

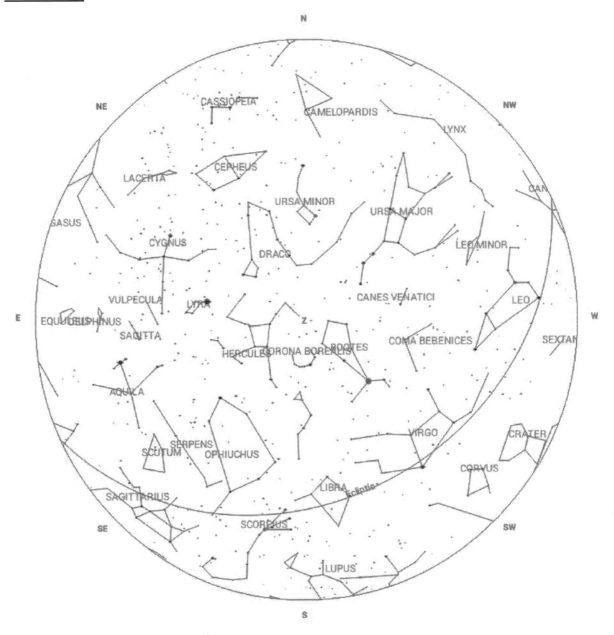

Designation	Name	Con.	Type	R.A.	Dec.	Mag	Size/Sep
Eps Boo	Izar	Boo	MS	14h 45m	+27° 04'	2.4	3"
Xi Boo		Boo	MS	14h 51m	+19° 06'	4.6	6"
Kap Boo	Asellus Tertius	Boo	MS	14h 14m	+51° 47'	4.5	13"
Pi Boo		Boo	MS	14h 41m	+16° 25'	4.5	6"
39 Boo		Boo	MS	14h 50m	+48° 43'	5.7	3"
Struve 1835		Boo	MS	14h 23m	+08° 27'	4.9	6"
Mu Boo	Alkalurops	Boo	MS	15h 24m	+37° 23'	4.3	108"
44 Boo		Boo	MS	15h 04m	+47° 39'	4.8	2"
Del Boo		Boo	MS	15h 16m	+33° 19'	3.5	105"

Designation	Name	Con.	Type	R.A.	Dec.	Mag	Size/Sep
Nu Boo		Boo	MS	15h 31m	+40° 50'	5.0	15'
Zet CrB		CrB	MS	15h 39m	+36° 38'	4.6	6"
Sig CrB		CrB	MS	16h 15m	+33° 52'	5.7	7"
R CrB	Fade Out Star	CrB	Var	15h 49m	+28° 09'	5.7-14.8	N/A
T CrB	Blaze Star	CrB	RN	16h 00m	+25° 55'	2.0-10.8	N/A
M 3		CVn	GC	13h 42m	+28° 23'	6.3	18'
16/17 Dra		Dra	MS	16h 36m	+52° 55'	5.1	90"
Nu Dra	Kuma	Dra	MS	17h 32m	+55° 10'	4.9	63"
Mu Dra		Dra	MS	17h 05m	+54° 28'	5.8	2"
Psi Dra		Dra	MS	17h 42m	+72° 09'	4.6	30"
39 Dra		Dra	MS	18h 24m	+58° 48'	5.0	89"
40/41 Dra		Dra	MS	18h 00m	+80° 00'	5.7	222"
M 13	Keystone Cluster	Her	GC	16h 42m	+36° 27'	5.8	20'
Kap Her	Marfik	Her	MS	16h 41m	+31° 36'	5.0	28"
M 92		Her	GC	17h 17m	+43° 08'	6.5	14'
Alp Her	Rasalgethi	Her	MS	17h 15m	+14° 23'	3.1	5"
Del Her	Sarin	Her	MS	17h 15m	+24° 50'	3.1	14"
Rho Her		Her	MS	17h 24m	+37° 09'	4.2	4"
95 Her		Her	MS	18h 02m	+21° 36'	4.3	6"
100 Her		Her	MS	18h 08m	+26° 06'	5.8	14"
Alp Lib	Zuben Elgenubi	Lib	MS	14h 51m	-16° 02'	2.8	230"
Bet Lib	The Emerald Star	Lib	*	15h 14m	-09° 23'	2.6	N/A
Struve 1962		Lib	MS	15h 39m	-08° 47'	5.4	12"
Eps Lyr	The Double Double	Lyr	MS	18h 44m	+39° 40'	4.7	3"
Bet Lyr	Sheliak	Lyr	MS	18h 50m	+33° 22'	3.2	86"
Del Lyr		Lyr	MS	18h 54m	+36° 58'	4.2	630"
Zet Lyr		Lyr	MS	18h 45m	+37° 36'	4.4	44"
IC 4665	Summer Beehive	Oph	OC	17h 46m	+05° 43'	5.3	70'
M 10		Oph	GC	16h 57m	-04° 06'	6.6	20'
M 12		Oph	GC	16h 47m	-01° 57'	6.1	16'
Rho Oph		Oph	MS	16h 26m	-23° 27'	4.6	3"
M 62		Oph	GC	17h 01m	-30° 07'	6.4	15'
36 Oph		Oph	MS	17h 15m	-26° 36'	4.3	730"
Omi Oph		Oph	MS	17h 18m	-24° 17'	5.1	10"
61 Oph		Oph	MS	17h 45m	+02° 35'	6.2	21"
NGC 6633	Tweedledum Cluster	Oph	OC	18h 27m	+06° 31'	5.6	20'
70 Oph		Oph	MS	18h 06m	+02° 30'	4.2	4"
IC 4756		Ser	OC	18h 39m	+05° 27'	5.4	39'
M 16	Eagle Nebula	Ser	Neb	18h 19m	-13° 49'	6.0	9'
The Ser		Ser	MS	18h 56m	+04° 12'	4.3	22"
M 5		Ser	GC	15h 19m	+02° 05'	5.7	23'
Del Ser		Ser	MS	15h 35m	+10° 32'	4.2	4"
Zet UMa	Mizar & Alcor	UMa	MS	13h 24m	+54° 56'	2.1	711"
The Vir		Vir	MS	13h 10m	-05° 32'	4.4	70"

Chart 17

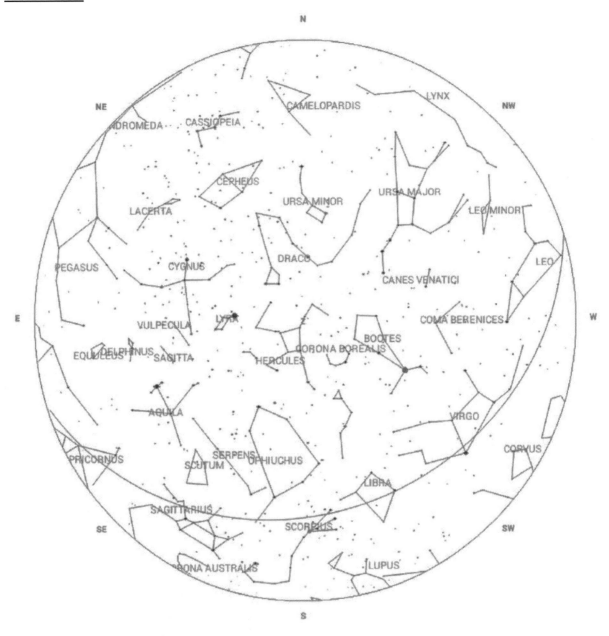

Designation	Name	Con.	Type	R.A.	Dec.	Mag	Size/Sep
15 Aql		Aql	MS	19h 06m	-04° 00'	5.4	39"
57 Aql		Aql	MS	19h 55m	-08° 14'	5.7	36"
Eps Boo	Izar	Boo	MS	14h 45m	+27° 04'	2.4	3"
Xi Boo		Boo	MS	14h 51m	+19° 06'	4.6	6"
Kap Boo	Asellus Tertius	Boo	MS	14h 14m	+51° 47'	4.5	13"
Pi Boo		Boo	MS	14h 41m	+16° 25'	4.5	6"
39 Boo		Boo	MS	14h 50m	+48° 43'	5.7	3"
Struve 1835		Boo	MS	14h 23m	+08° 27'	4.9	6"
Mu Boo	Alkalurops	Boo	MS	15h 24m	+37° 23'	4.3	108"

Designation	Name	Con.	Type	R.A.	Dec.	Mag	Size/Sep
44 Boo		Boo	MS	15h 04m	+47° 39'	4.8	2"
Del Boo		Boo	MS	15h 16m	+33° 19'	3.5	105"
Nu Boo		Boo	MS	15h 31m	+40° 50'	5.0	15'
Zet CrB		CrB	MS	15h 39m	+36° 38'	4.6	6"
Sig CrB		CrB	MS	16h 15m	+33° 52'	5.7	7"
Bet Cyg	Albireo	Cyg	MS	19h 31m	+27° 58'	3.1	34"
Del Cyg		Cyg	MS	19h 45m	+45° 08'	2.9	2"
16/17 Dra		Dra	MS	16h 36m	+52° 55'	5.1	90"
Nu Dra	Kuma	Dra	MS	17h 32m	+55° 10'	4.9	63"
Psi Dra		Dra	MS	17h 42m	+72° 09'	4.6	30"
39 Dra		Dra	MS	18h 24m	+58° 48'	5.0	89"
40/41 Dra		Dra	MS	18h 00m	+80° 00'	5.7	222"
Eps Dra		Dra	MS	19h 48m	+70° 16'	3.8	3"
Kap Her	Marfik	Her	MS	16h 41m	+31° 36'	5.0	28"
Alp Her	Rasalgethi	Her	MS	17h 15m	+14° 23'	3.1	5"
Del Her	Sarin	Her	MS	17h 15m	+24° 50'	3.1	14"
Rho Her		Her	MS	17h 24m	+37° 09'	4.2	4"
95 Her		Her	MS	18h 02m	+21° 36'	4.3	6"
Alp Lib	Zuben Elgenubi	Lib	MS	14h 51m	-16° 02'	2.8	230"
Bet Lib	The Emerald Star	Lib	*	15h 14m	-09° 23'	2.6	N/A
Struve 1962		Lib	MS	15h 39m	-08° 47'	5.4	12"
Eps Lyr	The Double Double	Lyr	MS	18h 44m	+39° 40'	4.7	3"
Bet Lyr	Sheliak	Lyr	MS	18h 50m	+33° 22'	3.2	86"
Del Lyr		Lyr	MS	18h 54m	+36° 58'	4.2	630"
Zet Lyr		Lyr	MS	18h 45m	+37° 36'	4.4	44"
IC 4665	Summer Beehive	Oph	OC	17h 46m	+05° 43'	5.3	70'
Rho Oph		Oph	MS	16h 26m	-23° 27'	4.6	3"
36 Oph		Oph	MS	17h 15m	-26° 36'	4.3	730"
Omi Oph		Oph	MS	17h 18m	-24° 17'	5.1	10"
NGC 6633	Tweedledum Cluster	Oph	OC	18h 27m	+06° 31'	5.6	20'
70 Oph		Oph	MS	18h 06m	+02° 30'	4.2	4"
M 6	Butterfly Cluster	Sco	OC	17h 40m	-32° 15'	4.6	20'
M 7		Sco	OC	17h 54m	-34° 48'	3.3	80'
M 4		Sco	GC	16h 24m	-26° 32'	5.4	36'
Alp Sco	Antares	Sco	MS	16h 29m	-26° 26'	1.0	3"
Bet Sco	Graffias	Sco	MS	16h 05m	-19° 48'	2.6	14"
Nu Sco	Jabbah	Sco	MS	16h 12m	-19° 28'	4.0	2"
Xi Sco		Sco	MS	16h 04m	-11° 22'	4.2	8"
Struve 1999		Sco	MS	16h 04m	-11° 22'	4.2	12"
IC 4756		Ser	OC	18h 39m	+05° 27'	5.4	39'
The Ser		Ser	MS	18h 56m	+04° 12'	4.3	22"
M 5		Ser	GC	15h 19m	+02° 05'	5.7	23'
Del Ser		Ser	MS	15h 35m	+10° 32'	4.2	4"
Collinder 399	Coathanger	Vul	Ast	19h 25m	+20° 11'	4.8	89'

Chart 18

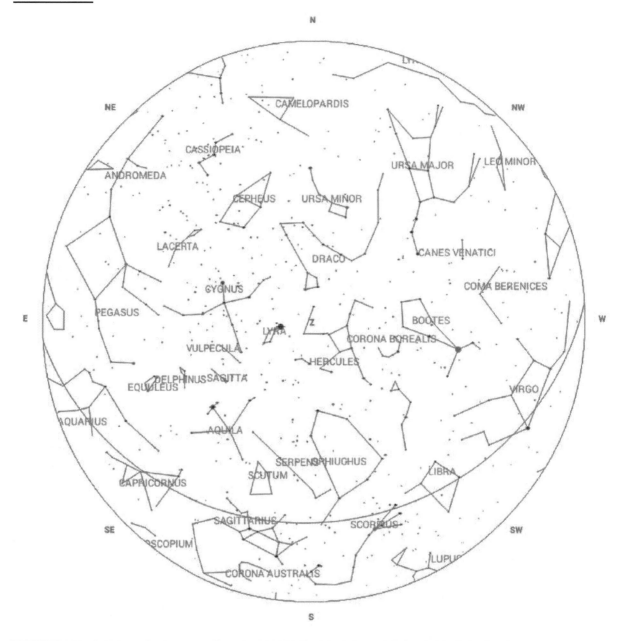

Designation	Name	Con.	Type	R.A.	Dec.	Mag	Size/Sep
15 Aql		Aql	MS	19h 06m	-04° 00'	5.4	39"
57 Aql		Aql	MS	19h 55m	-08° 14'	5.7	36"
Mu Boo	Alkalurops	Boo	MS	15h 24m	+37° 23'	4.3	108"
44 Boo		Boo	MS	15h 04m	+47° 39'	4.8	2"
Del Boo		Boo	MS	15h 16m	+33° 19'	3.5	105"
Nu Boo		Boo	MS	15h 31m	+40° 50'	5.0	15'
Zet CrB		CrB	MS	15h 39m	+36° 38'	4.6	6"
Sig CrB		CrB	MS	16h 15m	+33° 52'	5.7	7"
Bet Cyg	Albireo	Cyg	MS	19h 31m	+27° 58'	3.1	34"

Designation	Name	Con.	Type	R.A.	Dec.	Mag	Size/Sep
Del Cyg		Cyg	MS	19h 45m	+45° 08'	2.9	2"
NGC 7000	North American Nebula	Cyg	Neb	20h 59m	+44° 22'	4.0	120'
Gam Del		Del	MS	20h 47m	+16° 07'	3.9	9"
16/17 Dra		Dra	MS	16h 36m	+52° 55'	5.1	90"
Nu Dra	Kuma	Dra	MS	17h 32m	+55° 10'	4.9	63"
Psi Dra		Dra	MS	17h 42m	+72° 09'	4.6	30"
39 Dra		Dra	MS	18h 24m	+58° 48'	5.0	89"
40/41 Dra		Dra	MS	18h 00m	+80° 00'	5.7	222"
Eps Dra		Dra	MS	19h 48m	+70° 16'	3.8	3"
Eps Equ		Equ	MS	20h 59m	+04° 18'	5.2	11"
Kap Her	Marfik	Her	MS	16h 41m	+31° 36'	5.0	28"
Alp Her	Rasalgethi	Her	MS	17h 15m	+14° 23'	3.1	5"
Del Her	Sarin	Her	MS	17h 15m	+24° 50'	3.1	14"
Rho Her		Her	MS	17h 24m	+37° 09'	4.2	4"
95 Her		Her	MS	18h 02m	+21° 36'	4.3	6"
Eps Lyr	The Double Double	Lyr	MS	18h 44m	+39° 40'	4.7	3"
Bet Lyr	Sheliak	Lyr	MS	18h 50m	+33° 22'	3.2	86"
Del Lyr		Lyr	MS	18h 54m	+36° 58'	4.2	630"
Zet Lyr		Lyr	MS	18h 45m	+37° 36'	4.4	44"
IC 4665	Summer Beehive	Oph	OC	17h 46m	+05° 43'	5.3	70'
Rho Oph		Oph	MS	16h 26m	-23° 27'	4.6	3"
36 Oph		Oph	MS	17h 15m	-26° 36'	4.3	730"
Omi Oph		Oph	MS	17h 18m	-24° 17'	5.1	10"
NGC 6633	Tweedledum Cluster	Oph	OC	18h 27m	+06° 31'	5.6	20'
70 Oph		Oph	MS	18h 06m	+02° 30'	4.2	4"
M 6	Butterfly Cluster	Sco	OC	17h 40m	-32° 15'	4.6	20'
M 7		Sco	OC	17h 54m	-34° 48'	3.3	80'
M 4		Sco	GC	16h 24m	-26° 32'	5.4	36'
Alp Sco	Antares	Sco	MS	16h 29m	-26° 26'	1.0	3"
Bet Sco	Graffias	Sco	MS	16h 05m	-19° 48'	2.6	14"
Nu Sco	Jabbah	Sco	MS	16h 12m	-19° 28'	4.0	2"
Xi Sco		Sco	MS	16h 04m	-11° 22'	4.2	8"
Struve 1999		Sco	MS	16h 04m	-11° 22'	4.2	12"
IC 4756		Ser	OC	18h 39m	+05° 27'	5.4	39'
The Ser		Ser	MS	18h 56m	+04° 12'	4.3	22"
M 5		Ser	GC	15h 19m	+02° 05'	5.7	23'
Del Ser		Ser	MS	15h 35m	+10° 32'	4.2	4"
The Sge		Sge	MS	20h 10m	+20° 55'	4.6	84"
M 24	Sagittarius Star Cloud	Sgr	OC	18h 18m	-18° 24'	3.1	90'
M 22		Sgr	GC	18h 36m	-23° 54'	5.2	32'
M 8	Lagoon Nebula	Sgr	Neb	18h 04m	-24° 23'	5.0	17'
Collinder 399	Coathanger	Vul	Ast	19h 25m	+20° 11'	4.8	89'
NGC 6885		Vul	OC	20h 12m	+26° 29'	5.7	20'

Chart 19

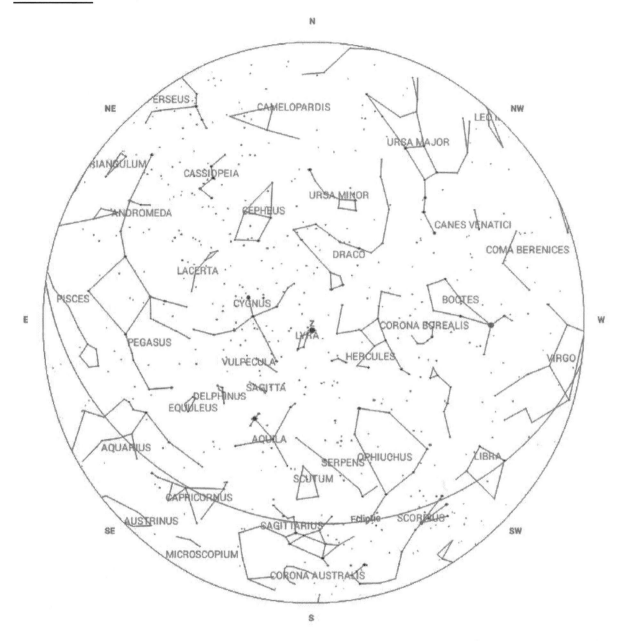

Designation	Name	Con.	Type	R.A.	Dec.	Mag	Size/Sep
15 Aql		Aql	MS	19h 06m	-04° 00'	5.4	39"
57 Aql		Aql	MS	19h 55m	-08° 14'	5.7	36"
Bet Cyg	Albireo	Cyg	MS	19h 31m	+27° 58'	3.1	34"
Del Cyg		Cyg	MS	19h 45m	+45° 08'	2.9	2"
16 Cyg		Cyg	MS	19h 42m	+50° 32'	5.9	39"
61 Cyg		Cyg	MS	21h 07m	+38° 44'	5.2	29"
M 39		Cyg	OC	21h 32m	+48° 26'	5.3	29'
Mu Cyg		Cyg	MS	21h 44m	+28° 45'	4.5	200"

Designation	Name	Con.	Type	R.A.	Dec.	Mag	Size/Sep
V460		Cyg	Var/CS	21h 42m	+35° 31'	5.6-7.0	N/A
NGC 7000	North American Nebula	Cyg	Neb	20h 59m	+44° 22'	4.0	120'
	Northern Coalsack	Cyg	DN	20h 41m	+43° 00'	6.0	60'
Gam Del		Del	MS	20h 47m	+16° 07'	3.9	9"
16/17 Dra		Dra	MS	16h 36m	+52° 55'	5.1	90"
Nu Dra	Kuma	Dra	MS	17h 32m	+55° 10'	4.9	63"
Mu Dra		Dra	MS	17h 05m	+54° 28'	5.8	2"
Psi Dra		Dra	MS	17h 42m	+72° 09'	4.6	30"
39 Dra		Dra	MS	18h 24m	+58° 48'	5.0	89"
40/41 Dra		Dra	MS	18h 00m	+80° 00'	5.7	222"
Eps Dra		Dra	MS	19h 48m	+70° 16'	3.8	3"
Eps Equ		Equ	MS	20h 59m	+04° 18'	5.2	11"
M 13	Keystone Cluster	Her	GC	16h 42m	+36° 27'	5.8	20'
Kap Her	Marfik	Her	MS	16h 41m	+31° 36'	5.0	28"
Alp Her	Rasalgethi	Her	MS	17h 15m	+14° 23'	3.1	5"
Del Her	Sarin	Her	MS	17h 15m	+24° 50'	3.1	14"
Rho Her		Her	MS	17h 24m	+37° 09'	4.2	4"
95 Her		Her	MS	18h 02m	+21° 36'	4.3	6"
100 Her		Her	MS	18h 08m	+26° 06'	5.8	14"
Eps Lyr	The Double Double	Lyr	MS	18h 44m	+39° 40'	4.7	3"
Bet Lyr	Sheliak	Lyr	MS	18h 50m	+33° 22'	3.2	86"
Del Lyr		Lyr	MS	18h 54m	+36° 58'	4.2	630"
Zet Lyr		Lyr	MS	18h 45m	+37° 36'	4.4	44"
IC 4665	Summer Beehive	Oph	OC	17h 46m	+05° 43'	5.3	70'
M 12		Oph	GC	16h 47m	-01° 57'	6.1	16'
Rho Oph		Oph	MS	16h 26m	-23° 27'	4.6	3"
36 Oph		Oph	MS	17h 15m	-26° 36'	4.3	730"
Omi Oph		Oph	MS	17h 18m	-24° 17'	5.1	10"
61 Oph		Oph	MS	17h 45m	+02° 35'	6.2	21"
NGC 6633	Tweedledum Cluster	Oph	OC	18h 27m	+06° 31'	5.6	20'
70 Oph		Oph	MS	18h 06m	+02° 30'	4.2	4"
IC 4756		Ser	OC	18h 39m	+05° 27'	5.4	39'
M 16	Eagle Nebula	Ser	Neb	18h 19m	-13° 49'	6.0	9'
The Ser		Ser	MS	18h 56m	+04° 12'	4.3	22"
The Sge		Sge	MS	20h 10m	+20° 55'	4.6	84"
15 Sge		Sge	MS	20h 04m	+17° 04'	5.8	204"
M 17	Swan Nebula	Sgr	Neb	18h 21m	-16° 11'	6.0	11'
M 25		Sgr	OC	18h 32m	-19° 07'	6.2	29'
M 24	Sagittarius Star Cloud	Sgr	OC	18h 18m	-18° 24'	3.1	90'
M 22		Sgr	GC	18h 36m	-23° 54'	5.2	32'
M 8	Lagoon Nebula	Sgr	Neb	18h 04m	-24° 23'	5.0	17'
M 23		Sgr	OC	17h 57m	-18° 59'	5.9	29'
Collinder 399	Coathanger	Vul	Ast	19h 25m	+20° 11'	4.8	89'
NGC 6885		Vul	OC	20h 12m	+26° 29'	5.7	20'

Chart 20

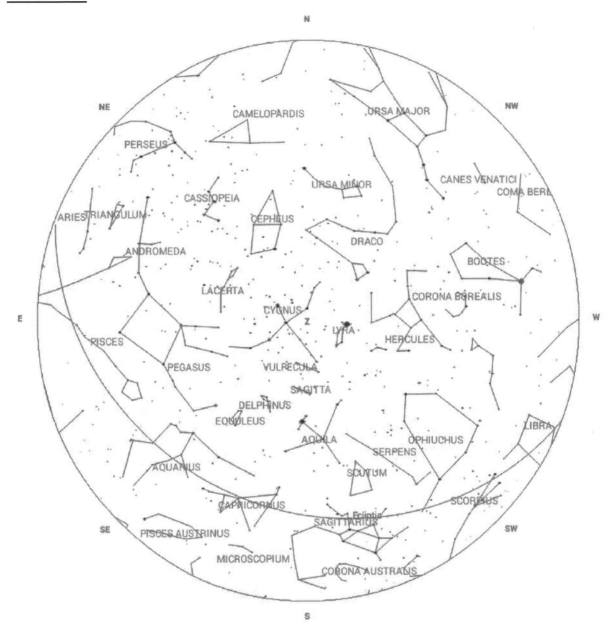

Designation	Name	Con.	Type	R.A.	Dec.	Mag	Size/Sep
15 Aql		Aql	MS	19h 06m	-04° 00'	5.4	39"
57 Aql		Aql	MS	19h 55m	-08° 14'	5.7	36"
Alp Cap	Al Giedi	Cap	MS	20h 18m	-12° 32'	3.6	45"
Bet Cap	Dabih	Cap	MS	20h 21m	-14° 47'	3.1	205"
Omi Cap		Cap	MS	20h 30m	-18° 35'	5.9	22"
IC 1396	Misty Clover Cluster	Cep	OC	21h 39m	+57° 30'	5.1	89'
Bet Cep	Alfirk	Cep	MS	21h 29m	+70° 34'	3.2	13"
Del Cep		Cep	MS/Var	22h 29m	+58° 25'	3.5-4.4	41"
Struve 2816		Cep	MS	21h 39m	+57° 29'	5.7	20"

Designation	Name	Con.	Type	R.A.	Dec.	Mag	Size/Sep
Mu Cep	Herschel's Garnet Star	Cep	Var/CS	21h 44m	+58° 47'	3.4-5.1	N/A
Struve 2840		Cep	MS	21h 52m	+55° 48'	5.7	18"
Xi Cep	Alkurhah	Cep	MS	22h 04m	+64° 38'	4.3	8"
Bet Cyg	Albireo	Cyg	MS	19h 31m	+27° 58'	3.1	34"
Del Cyg		Cyg	MS	19h 45m	+45° 08'	2.9	2"
16 Cyg		Cyg	MS	19h 42m	+50° 32'	5.9	39"
61 Cyg		Cyg	MS	21h 07m	+38° 44'	5.2	29"
M 39		Cyg	OC	21h 32m	+48° 26'	5.3	29'
Mu Cyg		Cyg	MS	21h 44m	+28° 45'	4.5	200"
NGC 7000	North American Nebula	Cyg	Neb	20h 59m	+44° 22'	4.0	120'
Gam Del		Del	MS	20h 47m	+16° 07'	3.9	9"
Nu Dra	Kuma	Dra	MS	17h 32m	+55° 10'	4.9	63"
Mu Dra		Dra	MS	17h 05m	+54° 28'	5.8	2"
Psi Dra		Dra	MS	17h 42m	+72° 09'	4.6	30"
39 Dra		Dra	MS	18h 24m	+58° 48'	5.0	89"
40/41 Dra		Dra	MS	18h 00m	+80° 00'	5.7	222"
Eps Dra		Dra	MS	19h 48m	+70° 16'	3.8	3"
Eps Equ		Equ	MS	20h 59m	+04° 18'	5.2	11"
Alp Her	Rasalgethi	Her	MS	17h 15m	+14° 23'	3.1	5"
Del Her	Sarin	Her	MS	17h 15m	+24° 50'	3.1	14"
Rho Her		Her	MS	17h 24m	+37° 09'	4.2	4"
95 Her		Her	MS	18h 02m	+21° 36'	4.3	6"
100 Her		Her	MS	18h 08m	+26° 06'	5.8	14"
8 Lac		Lac	MS	22h 36m	+39° 38'	5.7	82"
Eps Lyr	The Double Double	Lyr	MS	18h 44m	+39° 40'	4.7	3"
Bet Lyr	Sheliak	Lyr	MS	18h 50m	+33° 22'	3.2	86"
Del Lyr		Lyr	MS	18h 54m	+36° 58'	4.2	630"
Zet Lyr		Lyr	MS	18h 45m	+37° 36'	4.4	44"
IC 4665	Summer Beehive	Oph	OC	17h 46m	+05° 43'	5.3	70'
36 Oph		Oph	MS	17h 15m	-26° 36'	4.3	730"
Omi Oph		Oph	MS	17h 18m	-24° 17'	5.1	10"
NGC 6633	Tweedledum Cluster	Oph	OC	18h 27m	+06° 31'	5.6	20'
70 Oph		Oph	MS	18h 06m	+02° 30'	4.2	4"
Eps Peg	Enif	Peg	MS	21h 44m	+09° 52'	2.1	143"
IC 4756		Ser	OC	18h 39m	+05° 27'	5.4	39'
The Ser		Ser	MS	18h 56m	+04° 12'	4.3	22"
The Sge		Sge	MS	20h 10m	+20° 55'	4.6	84"
15 Sge		Sge	MS	20h 04m	+17° 04'	5.8	204"
M 24	Sagittarius Star Cloud	Sgr	OC	18h 18m	-18° 24'	3.1	90'
M 22		Sgr	GC	18h 36m	-23° 54'	5.2	32'
M 8	Lagoon Nebula	Sgr	Neb	18h 04m	-24° 23'	5.0	17'
M 23		Sgr	OC	17h 57m	-18° 59'	5.9	29'
Collinder 399	Coathanger	Vul	Ast	19h 25m	+20° 11'	4.8	89'
NGC 6885		Vul	OC	20h 12m	+26° 29'	5.7	20'

Chart 21

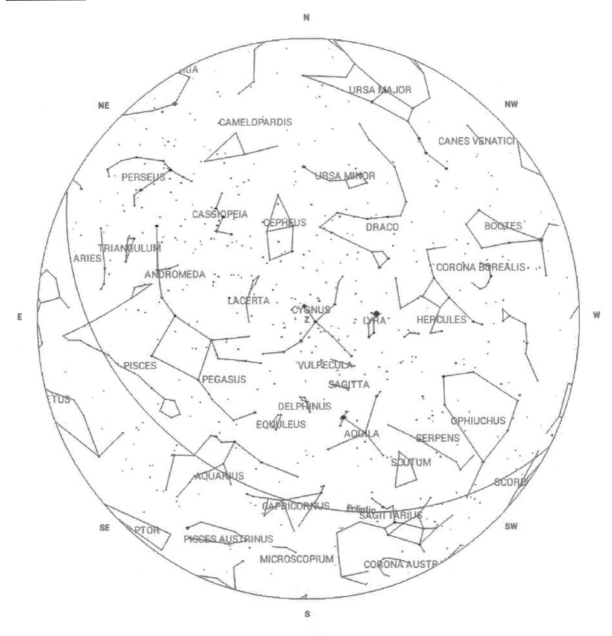

Designation	Name	Con.	Type	R.A.	Dec.	Mag	Size/Sep
Struve 2404		Aql	MS	18h 51m	+10° 59'	6.4	4"
15 Aql		Aql	MS	19h 06m	-04° 00'	5.4	39"
57 Aql		Aql	MS	19h 55m	-08° 14'	5.7	36"
V Aql		Aql	Var	19h 04m	-05° 41'	6.6-8.4	N/A
107 Aqr		Aqr	MS	23h 47m	-18° 35'	5.3	7"
94 Aqr		Aqr	MS	23h 19m	-13° 28'	5.2	13"
NGC 7293	Helix Nebula	Aqr	PN	22h 30m	-20° 50'	6.3	16'
Zet Aqr		Aqr	MS	22h 29m	-00° 01'	3.7	2"
41 Aqr		Aqr	MS	22h 14m	-21° 04'	5.3	5"

Designation	Name	Con.	Type	R.A.	Dec.	Mag	Size/Sep
53 Aqr		Aqr	MS	22h 27m	-16° 45'	5.6	3"
M 2		Aqr	GC	21h 33m	-00° 49'	6.6	16'
Alp Cap	Al Giedi	Cap	MS	20h 18m	-12° 32'	3.6	45"
Bet Cap	Dabih	Cap	MS	20h 21m	-14° 47'	3.1	205"
RT Cap		Cap	Var/CS	20h 17m	-21° 19'	6.5-8.1	N/A
Omi Cap		Cap	MS	20h 30m	-18° 35'	5.9	22"
Sig Cas		Cas	MS	23h 59m	+55° 45'	4.9	3"
IC 1396	Misty Clover Cluster	Cep	OC	21h 39m	+57° 30'	5.1	89'
Bet Cep	Alfirk	Cep	MS	21h 29m	+70° 34'	3.2	13"
Del Cep		Cep	MS/Var	22h 29m	+58° 25'	3.5-4.4	41"
Struve 2816		Cep	MS	21h 39m	+57° 29'	5.7	20"
Mu Cep	Herschel's Garnet Star	Cep	Var/CS	21h 44m	+58° 47'	3.4-5.1	N/A
Struve 2840		Cep	MS	21h 52m	+55° 48'	5.7	18"
Xi Cep	Alkurhah	Cep	MS	22h 04m	+64° 38'	4.3	8"
Omi Cep		Cep	MS	23h 19m	+68° 07'	4.8	3"
Bet Cyg	Albireo	Cyg	MS	19h 31m	+27° 58'	3.1	34"
Del Cyg		Cyg	MS	19h 45m	+45° 08'	2.9	2"
16 Cyg		Cyg	MS	19h 42m	+50° 32'	5.9	39"
61 Cyg		Cyg	MS	21h 07m	+38° 44'	5.2	29"
M 39		Cyg	OC	21h 32m	+48° 26'	5.3	29'
Mu Cyg		Cyg	MS	21h 44m	+28° 45'	4.5	200"
V460		Cyg	Var/CS	21h 42m	+35° 31'	5.6-7.0	N/A
NGC 7000	North American Nebula	Cyg	Neb	20h 59m	+44° 22'	4.0	120'
	Northern Coalsack	Cyg	DN	20h 41m	+43° 00'	6.0	60'
Gam Del		Del	MS	20h 47m	+16° 07'	3.9	9"
39 Dra		Dra	MS	18h 24m	+58° 48'	5.0	89"
40/41 Dra		Dra	MS	18h 00m	+80° 00'	5.7	222"
Eps Dra		Dra	MS	19h 48m	+70° 16'	3.8	3"
UX Dra		Dra	Var/CS	19h 22m	+76° 34'	5.9-7.1	N/A
Eps Equ		Equ	MS	20h 59m	+04° 18'	5.2	11"
95 Her		Her	MS	18h 02m	+21° 36'	4.3	6"
100 Her		Her	MS	18h 08m	+26° 06'	5.8	14"
8 Lac		Lac	MS	22h 36m	+39° 38'	5.7	82"
Eps Lyr	The Double Double	Lyr	MS	18h 44m	+39° 40'	4.7	3"
Bet Lyr	Sheliak	Lyr	MS	18h 50m	+33° 22'	3.2	86"
Del Lyr		Lyr	MS	18h 54m	+36° 58'	4.2	630"
Zet Lyr		Lyr	MS	18h 45m	+37° 36'	4.4	44"
M 15		Peg	GC	21h 30m	+12° 10'	6.3	18'
Eps Peg	Enif	Peg	MS	21h 44m	+09° 52'	2.1	143"
M 11	Wild Duck Cluster	Sct	OC	18h 51m	-06° 16'	6.1	32'
The Sge		Sge	MS	20h 10m	+20° 55'	4.6	84"
15 Sge		Sge	MS	20h 04m	+17° 04'	5.8	204"
Collinder 399	Coathanger	Vul	Ast	19h 25m	+20° 11'	4.8	89'
NGC 6885		Vul	OC	20h 12m	+26° 29'	5.7	20'

Chart 22

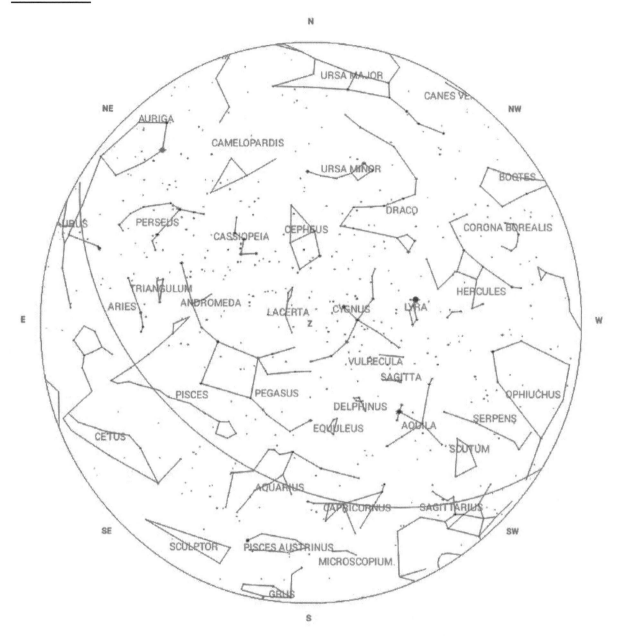

Designation	Name	Con.	Type	R.A.	Dec.	Mag	Size/Sep
M 31	Andromeda Galaxy	And	Gx	00h 43m	+41° 16'	4.3	156'
Pi And		And	MS	00h 38m	+33° 49'	4.4	36"
15 Aql		Aql	MS	19h 06m	-04° 00'	5.4	39"
57 Aql		Aql	MS	19h 55m	-08° 14'	5.7	36"
107 Aqr		Aqr	MS	23h 47m	-18° 35'	5.3	7"
94 Aqr		Aqr	MS	23h 19m	-13° 28'	5.2	13"
NGC 7293	Helix Nebula	Aqr	PN	22h 30m	-20° 50'	6.3	16'
Zet Aqr		Aqr	MS	22h 29m	-00° 01'	3.7	2"

Designation	Name	Con.	Type	R.A.	Dec.	Mag	Size/Sep
41 Aqr		Aqr	MS	22h 14m	-21° 04'	5.3	5"
53 Aqr		Aqr	MS	22h 27m	-16° 45'	5.6	3"
M 2		Aqr	GC	21h 33m	-00° 49'	6.6	16'
M 30		Cap	GC	21h 40m	-23° 11'	6.9	12'
Alp Cap	Al Giedi	Cap	MS	20h 18m	-12° 32'	3.6	45"
Bet Cap	Dabih	Cap	MS	20h 21m	-14° 47'	3.1	205"
Omi Cap		Cap	MS	20h 30m	-18° 35'	5.9	22"
Struve 3053		Cas	MS	00h 03m	+66° 06'	5.9	15"
Eta Cas	Achird	Cas	MS	00h 50m	+57° 54'	3.6	13"
Sig Cas		Cas	MS	23h 59m	+55° 45'	4.9	3"
IC 1396	Misty Clover Cluster	Cep	OC	21h 39m	+57° 30'	5.1	89'
Bet Cep	Alfirk	Cep	MS	21h 29m	+70° 34'	3.2	13"
Del Cep		Cep	MS/Var	22h 29m	+58° 25'	3.5-4.4	41"
Struve 2816		Cep	MS	21h 39m	+57° 29'	5.7	20"
Mu Cep	Herschel's Garnet Star	Cep	Var/CS	21h 44m	+58° 47'	3.4-5.1	N/A
Struve 2840		Cep	MS	21h 52m	+55° 48'	5.7	18"
Xi Cep	Alkurhah	Cep	MS	22h 04m	+64° 38'	4.3	8"
Omi Cep		Cep	MS	23h 19m	+68° 07'	4.8	3"
Bet Cyg	Albireo	Cyg	MS	19h 31m	+27° 58'	3.1	34"
Del Cyg		Cyg	MS	19h 45m	+45° 08'	2.9	2"
16 Cyg		Cyg	MS	19h 42m	+50° 32'	5.9	39"
NGC 6960	Veil Nebula (West)	Cyg	SNR	20h 46m	+30° 43'	7.0	63'
61 Cyg		Cyg	MS	21h 07m	+38° 44'	5.2	29"
M 39		Cyg	OC	21h 32m	+48° 26'	5.3	29'
Mu Cyg		Cyg	MS	21h 44m	+28° 45'	4.5	200"
V460		Cyg	Var/CS	21h 42m	+35° 31'	5.6-7.0	N/A
NGC 6992	Veil Nebula (East)	Cyg	SNR	20h 56m	+31° 43'	7.0	60'
NGC 7000	North American Nebula	Cyg	Neb	20h 59m	+44° 22'	4.0	120'
	Northern Coalsack	Cyg	DN	20h 41m	+43° 00'	6.0	60'
Gam Del		Del	MS	20h 47m	+16° 07'	3.9	9"
Eps Dra		Dra	MS	19h 48m	+70° 16'	3.8	3"
Eps Equ		Equ	MS	20h 59m	+04° 18'	5.2	11"
NGC 7243		Lac	OC	22h 15m	+49° 54'	6.7	29'
8 Lac		Lac	MS	22h 36m	+39° 38'	5.7	82"
Struve 2470/2474	The Double Double's Double	Lyr	MS	19h 09m	+34° 41'	6.7	16"
M 15		Peg	GC	21h 30m	+12° 10'	6.3	18'
Eps Peg	Enif	Peg	MS	21h 44m	+09° 52'	2.1	143"
TX Psc		Psc	Var/CS	23h 46m	+03° 29'	4.5-5.3	N/A
55 Psc		Psc	MS	00h 40m	+21° 26'	5.4	6"
65 Psc		Psc	MS	00h 50m	+27° 43'	7.0	4"
The Sge		Sge	MS	20h 10m	+20° 55'	4.6	84"
15 Sge		Sge	MS	20h 04m	+17° 04'	5.8	204"
Collinder 399	Coathanger	Vul	Ast	19h 25m	+20° 11'	4.8	89'
NGC 6885		Vul	OC	20h 12m	+26° 29'	5.7	20'

Chart 23

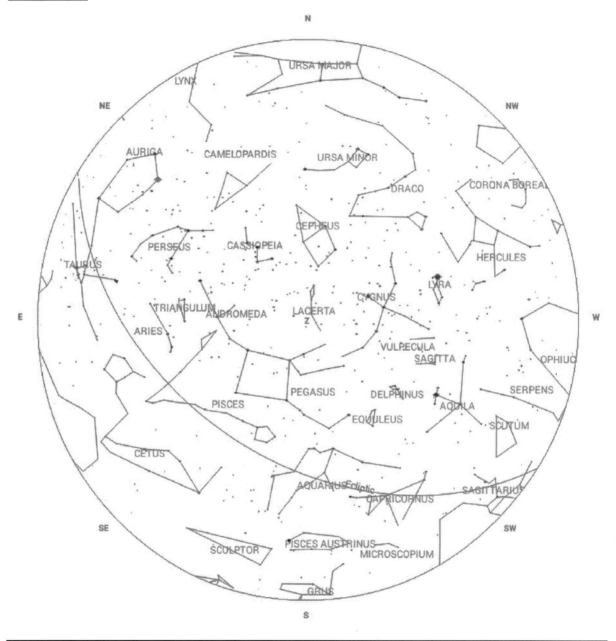

Designation	Name	Con.	Type	R.A.	Dec.	Mag	Size/Sep
M 31	Andromeda Galaxy	And	Gx	00h 43m	+41° 16'	4.3	156'
NGC 752	Golf Ball Cluster	And	OC	01h 58m	+37° 47'	6.6	75'
Pi And		And	MS	00h 38m	+33° 49'	4.4	36"
107 Aqr		Aqr	MS	23h 47m	-18° 35'	5.3	7"
94 Aqr		Aqr	MS	23h 19m	-13° 28'	5.2	13"
NGC 7293	Helix Nebula	Aqr	PN	22h 30m	-20° 50'	6.3	16'
Zet Aqr		Aqr	MS	22h 29m	-00° 01'	3.7	2"
41 Aqr		Aqr	MS	22h 14m	-21° 04'	5.3	5"

Designation	Name	Con.	Type	R.A.	Dec.	Mag	Size/Sep
53 Aqr		Aqr	MS	22h 27m	-16° 45'	5.6	3"
M 2		Aqr	GC	21h 33m	-00° 49'	6.6	16'
Lam Ari		Ari	MS	01h 59m	+23° 41'	4.8	37"
Gam Ari	Mesarthim	Ari	MS	01h 54m	+19° 22'	4.6	8"
M 30		Cap	GC	21h 40m	-23° 11'	6.9	12'
Alp Cap	Al Giedi	Cap	MS	20h 18m	-12° 32'	3.6	45"
Bet Cap	Dabih	Cap	MS	20h 21m	-14° 47'	3.1	205"
Omi Cap		Cap	MS	20h 30m	-18° 35'	5.9	22"
NGC 457	Owl Cluster	Cas	OC	01h 20m	+58° 17'	5.1	20'
Struve 163		Cas	MS	01h 51m	+64° 51'	6.5	35"
M 103		Cas	OC	01h 33m	+60° 39'	6.9	5'
Struve 3053		Cas	MS	00h 03m	+66° 06'	5.9	15"
Eta Cas	Achird	Cas	MS	00h 50m	+57° 54'	3.6	13"
Sig Cas		Cas	MS	23h 59m	+55° 45'	4.9	3"
NGC 663		Cas	OC	01h 46m	+61° 14'	6.4	14'
IC 1396	Misty Clover Cluster	Cep	OC	21h 39m	+57° 30'	5.1	89'
Bet Cep	Alfirk	Cep	MS	21h 29m	+70° 34'	3.2	13"
Del Cep		Cep	MS/Var	22h 29m	+58° 25'	3.5-4.4	41"
Struve 2816		Cep	MS	21h 39m	+57° 29'	5.7	20"
Mu Cep	Herschel's Garnet Star	Cep	Var/CS	21h 44m	+58° 47'	3.4-5.1	N/A
Struve 2840		Cep	MS	21h 52m	+55° 48'	5.7	18"
Xi Cep	Alkurhah	Cep	MS	22h 04m	+64° 38'	4.3	8"
Omi Cep		Cep	MS	23h 19m	+68° 07'	4.8	3"
NGC 6960	Veil Nebula (West)	Cyg	SNR	20h 46m	+30° 43'	7.0	63'
61 Cyg		Cyg	MS	21h 07m	+38° 44'	5.2	29"
M 39		Cyg	OC	21h 32m	+48° 26'	5.3	29'
Mu Cyg		Cyg	MS	21h 44m	+28° 45'	4.5	200"
NGC 6992	Veil Nebula (East)	Cyg	SNR	20h 56m	+31° 43'	7.0	60'
NGC 7000	North American Nebula	Cyg	Neb	20h 59m	+44° 22'	4.0	120'
	Northern Coalsack	Cyg	DN	20h 41m	+43° 00'	6.0	60'
Gam Del		Del	MS	20h 47m	+16° 07'	3.9	9"
Eps Equ		Equ	MS	20h 59m	+04° 18'	5.2	11"
NGC 7243		Lac	OC	22h 15m	+49° 54'	6.7	29'
8 Lac		Lac	MS	22h 36m	+39° 38'	5.7	82"
M 15		Peg	GC	21h 30m	+12° 10'	6.3	18'
Eps Peg	Enif	Peg	MS	21h 44m	+09° 52'	2.1	143"
55 Psc		Psc	MS	00h 40m	+21° 26'	5.4	6"
65 Psc		Psc	MS	00h 50m	+27° 43'	7.0	4"
Psi1 Psc		Psc	MS	01h 06m	+21° 28'	5.3	30"
Zet Psc		Psc	MS	01h 14m	+07° 35'	5.2	23"
The Sge		Sge	MS	20h 10m	+20° 55'	4.6	84"
15 Sge		Sge	MS	20h 04m	+17° 04'	5.8	204"
M 33	Triangulum Galaxy	Tri	Gx	01h 34m	+30° 40'	6.4	62'
NGC 6885		Vul	OC	20h 12m	+26° 29'	5.7	20'

Chart 24

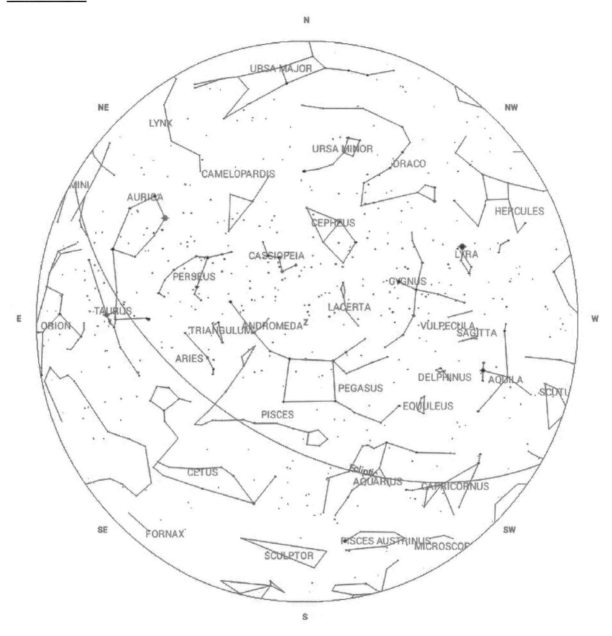

Designation	Name	Con.	Type	R.A.	Dec.	Mag	Size/Sep
Gam And	Almach	And	MS	02h 04m	+42° 20'	2.1	10"
M 31	Andromeda Galaxy	And	Gx	00h 43m	+41° 16'	4.3	156'
NGC 752	Golf Ball Cluster	And	OC	01h 58m	+37° 47'	6.6	75'
Pi And		And	MS	00h 38m	+33° 49'	4.4	36"
107 Aqr		Aqr	MS	23h 47m	-18° 35'	5.3	7"
94 Aqr		Aqr	MS	23h 19m	-13° 28'	5.2	13"
NGC 7293	Helix Nebula	Aqr	PN	22h 30m	-20° 50'	6.3	16'
Zet Aqr		Aqr	MS	22h 29m	-00° 01'	3.7	2"

Designation	Name	Con.	Type	R.A.	Dec.	Mag	Size/Sep
41 Aqr		Aqr	MS	22h 14m	-21° 04'	5.3	5"
53 Aqr		Aqr	MS	22h 27m	-16° 45'	5.6	3"
M 2		Aqr	GC	21h 33m	-00° 49'	6.6	16'
Lam Ari		Ari	MS	01h 59m	+23° 41'	4.8	37"
Gam Ari	Mesarthim	Ari	MS	01h 54m	+19° 22'	4.6	8"
30 Ari		Ari	MS	02h 37m	+24° 39'	6.5	39"
M 30		Cap	GC	21h 40m	-23° 11'	6.9	12'
NGC 457	Owl Cluster	Cas	OC	01h 20m	+58° 17'	5.1	20'
Struve 163		Cas	MS	01h 51m	+64° 51'	6.5	35"
M 103		Cas	OC	01h 33m	+60° 39'	6.9	5'
Struve 3053		Cas	MS	00h 03m	+66° 06'	5.9	15"
Eta Cas	Achird	Cas	MS	00h 50m	+57° 54'	3.6	13"
Sig Cas		Cas	MS	23h 59m	+55° 45'	4.9	3"
Iot Cas		Cas	MS	02h 29m	+67° 24'	4.5	7"
NGC 659	Ying Yang Cluster	Cas	OC	01h 44m	+60° 40'	7.2	5'
NGC 663		Cas	OC	01h 46m	+61° 14'	6.4	14'
IC 1396	Misty Clover Cluster	Cep	OC	21h 39m	+57° 30'	5.1	89'
Bet Cep	Alfirk	Cep	MS	21h 29m	+70° 34'	3.2	13"
Del Cep		Cep	MS/Var	22h 29m	+58° 25'	3.5-4.4	41"
Struve 2816		Cep	MS	21h 39m	+57° 29'	5.7	20"
Mu Cep	Herschel's Garnet Star	Cep	Var/CS	21h 44m	+58° 47'	3.4-5.1	N/A
Struve 2840		Cep	MS	21h 52m	+55° 48'	5.7	18"
Xi Cep	Alkurhah	Cep	MS	22h 04m	+64° 38'	4.3	8"
Omi Cep		Cep	MS	23h 19m	+68° 07'	4.8	3"
Gam Cet	Kaffajidhma	Cet	MS	02h 43m	+03° 14'	3.5	3"
Omi Cet	Mira	Cet	Var	02h 19m	-02° 59'	2.0-10.1	N/A
61 Cyg		Cyg	MS	21h 07m	+38° 44'	5.2	29"
M 39		Cyg	OC	21h 32m	+48° 26'	5.3	29'
Mu Cyg		Cyg	MS	21h 44m	+28° 45'	4.5	200"
V460		Cyg	Var/CS	21h 42m	+35° 31'	5.6-7.0	N/A
NGC 7243		Lac	OC	22h 15m	+49° 54'	6.7	29'
8 Lac		Lac	MS	22h 36m	+39° 38'	5.7	82"
M 34		Per	OC	02h 42m	+42° 46'	5.8	35'
NGC 869/884	Double Cluster	Per	OC	02h 21m	+57° 08'	4.4	18'
Eta Per		Per	MS	02h 51m	+55° 54'	3.8	29"
M 15		Peg	GC	21h 30m	+12° 10'	6.3	18'
Eps Peg	Enif	Peg	MS	21h 44m	+09° 52'	2.1	143"
Alp Psc	Alrisha	Psc	MS	02h 02m	+02° 46'	3.8	2"
TX Psc		Psc	Var/CS	23h 46m	+03° 29'	4.5-5.3	N/A
55 Psc		Psc	MS	00h 40m	+21° 26'	5.4	6"
Psi1 Psc		Psc	MS	01h 06m	+21° 28'	5.3	30"
Zet Psc		Psc	MS	01h 14m	+07° 35'	5.2	23"
M 33	Triangulum Galaxy	Tri	Gx	01h 34m	+30° 40'	6.4	62'
Alp UMi	Polaris	UMi	MS	02h 51m	+89° 20'	2.0	18"

2020 Summary Tables

Planet Visibility Ratings

		Morning Sky							Evening Sky						
		Me	Ve	Ma	Ju	Sa	Ur	Ne	Me	Ve	Ma	Ju	Sa	Ur	Ne
Jan	5th			*						**				***	**
	15th			*	*					**				**	**
	25th			*	*					**				**	**
Feb	5th			*	*	**			**	**				***	**
	15th			*	*	**				***				**	*
	25th			*	**	**				***				*	
Mar	5th	***		*	**	**				***				*	
	15th	****		*	**	**				***				*	
	25th	***		*	**	**				***				*	
Apr	5th	***		**	**	**		*		***					
	15th	**		**	***	***		*		****					
	25th			**	***	***		**		****					
May	5th			**	***	***		**		****					
	15th			**	****	***	*	**		***					
	25th			**	****	***	*	**	***						
Jun	5th			**	****	****	*	***	****						
	15th		***	**	****	****	*	***	***						
	25th		***	***	****	****	*	***							
Jul	5th		****	***	*****	****	**	****							
	15th	***	****	***		*****	**	****				*****			
	25th	***	****	***			**	****				*****	*****		
Aug	5th		****	****			**	****				*****	****		
	15th		***	****			***	*****				****	****		
	25th		***	****			***	*****				****	****		
Sep	5th		***	****			***	****				****	****		
	15th		***	*****			***		**			****	****		*****
	25th		**	*****			***		***			***	***		*****
Oct	5th		**	*****			***		***			***	***		*****
	15th		**				****		***	*****		***	***		****
	25th		**				****			*****		***	***		****
Nov	5th	**	**							****		**	**	****	****
	15th	**	**							****		**	**	****	****
	25th		**							****		**	**	***	***
Dec	5th		*							***		**	**	***	***
	15th		*							***		**	**	***	***
	25th		*							***		*	**	***	**

Solar and Lunar Eclipses

Date	Time (UT)	Type	Visible From
Jan 10th	19:09	Penumbral Lunar	Africa, Asia, Australia, Europe and north-eastern and northern North America.
Jun 5th	19:24	Penumbral Lunar	Africa, Asia, Australia, Europe and eastern South America.
Jun 21st	06:41	Annular Solar	Eastern Africa, Asia and the Indian Ocean.
Jul 5th	04:29	Penumbral Lunar	Africa, Central America, western Europe, North America and South America.
Nov 30th	09:43	Penumbral Lunar	Australia, western Europe, North America and South America.
Dec 14th	16:15	Total Solar	Antarctica, the southern Atlantic, the southern Pacific and South America.

Planetary Highlights

Date	Time (UT)	Elon.	Vis.	Description
Jan 16th	17:45	50° W	AM	Mars is 4.8° north of the bright star Antares. (Scorpius)
Jan 27th	19:25	39° E	PM	Venus is 0.1° south of Neptune. (Aquarius)
Mar 9th	14:32	42° E	PM	Venus is 2.4° north of Uranus. (Aries)
Mar 20th	06:14	66° W	AM	Mars is 0.7° south of Jupiter. (Sagittarius)
Mar 24th	02:03	25° W	AM	Mercury is at greatest western elongation from the Sun. (Aquarius)
Mar 24th	22:00	43° E	PM	Venus is at greatest eastern elongation from the Sun. (Aries)
Mar 31st	10:50	67° W	AM	Mars is 0.9° south of Saturn. (Capricornus)
Apr 3rd	11:59	44° W	PM	Venus is 0.3° south of the Pleiades star cluster. (Taurus.)
May 21st	07:49	19° E	PM	Mercury is 0.9° south of Venus. (Taurus)
Jun 4th	13:04	26° E	PM	Mercury is at greatest eastern elongation from the Sun. (Gemini)
Jun 12th	12:20	89° W	AM	Mars is 1.7° south of Neptune.(Aquarius)
Jul 11th	17:51	42° W	AM	Venus is 1.0° north of the bright star Aldebaran. (Taurus.)
Jul 14th	09:06	180°	AN	Jupiter is at opposition. (Sagittarius)
Jul 20th	23:33	180°	AN	Saturn is at opposition. (Sagittarius)
Aug 12th	23:49	48° W	AM	Venus is at greatest western elongation from the Sun. (Orion)
Sep 12th	09:20	180°	AN	Neptune is at opposition. (Aquarius)
Oct 1st	16:00	23° E	PM	Mercury is at greatest eastern elongation from the Sun. (Virgo)
Oct 2nd	17:50	37° W	AM	Venus is 0.1° south of the bright star Regulus. (Leo.)
Oct 14th	21:17	180°	AN	Mars is at opposition. (Pisces)
Oct 31st	17:15	180°	AN	Uranus is at opposition. (Aries)
Dec 21st	13:46	29° E	PM	Jupiter is 0.1° south of Saturn. (Capricornus)

Major Meteor Showers

Shower Name	Start Date	End Date	Peak	ZHR	Speed	Brightness	Moon
Quadrantids	Dec 28th	Jan 12th	Jan 3rd	120	***	*****	◑
Lyrids	Apr 18th	Apr 25th	Apr 22nd	18	***	*****	●
Eta Aquariids	Apr 24th	May 19th	May 7th	40	*	****	○
June Bootids	Jun 23rd	Jun 25th	Jun 24th	Var	*****	*****	◐
Alpha Capricornids	Jul 8th	Aug 10th	Jul 27th	5	*****	****	◐
Southern Delta Aquariids	Jul 21st	Aug 23rd	Jul 30th	16	***	*	○
Perseids	Jul 13th	Aug 26th	Aug 12th	100	*	*****	◕
Kappa Cygnids	Aug 6th	Aug 31st	Aug 17th	3	*****	**	●
Aurigids	Aug 29th	Sep 4th	Sep 1st	6	*	****	○
September Epsilon Perseids	Sep 5th	Sep 28th	Sep 9th	5	*	**	◑
Draconids	Oct 6th	Oct 10th	Oct 8th	Var	*****	***	◑
Southern Taurids	Sep 7th	Nov 19th	Oct 10th	5	****	****	◑
Orionids	Aug 25th	Nov 19th	Oct 22nd	15	*	****	◐
Andromedids	Oct 26th	Nov 20th	Nov 8th	Var	*****	****	◑
Northern Taurids	Oct 25th	Dec 4th	Nov 11th	5	****	****	◕
Leonids	Nov 5th	Dec 3rd	Nov 18th	15	*	****	◐
Alpha Monocerotids	Nov 21st	Nov 23rd	Nov 21st	Var	*	****	◑
Geminids	Nov 30th	Dec 17th	Dec 13th	120	****	***	●
December Leonis Minorids	Dec 6th	Jan 18th	Dec 20th	5	*	**	◐
Ursids	Dec 17th	Dec 24th	Dec 22nd	10	****	**	◐
Coma Berenicids	Dec 24th	Jan 3rd	Dec 31st	5	*	**	○

Key to the Monthly Guide Lunar and Planetary Data Tables

The table below provides a key to the abbreviations and acronyms used in the data tables for the Moon and planets. Please note that the position data (ie, the R.A. and declination) are accurate for 12:00 Universal Time. A complete list of all 88 constellations and their abbreviations can be found on the following pages.

+Cr	Waxing Crescent Moon	FQ	First Quarter Moon	LQ	Last Quarter Moon
-Cr	Waning Crescent Moon	+G	Waxing Gibbous Moon	NM	New Moon
FM	Full Moon	-G	Waning Gibbous Moon		
AM	Morning visibility	Elong	Elongation from Sun	Mag	Magnitude
Con	Constellation	Ill	Illumination	PM	Evening Sky
Dec	Declination	Mag	Magnitude	R.A.	Right Ascension
Diam	Apparent Diameter				

About the Events

In keeping with the standard used throughout the astronomical community, all of the times are shown in Universal Time and with that in mind, I've included a conversion table below. Also, bear in mind that if an event is not visible at the specific time it occurs (e.g., the Moon appearing close to a planet) it will always be worth observing at the nearest convenient opportunity.

You'll also notice that I don't provide angular separation for conjunctions involving the Moon. Since the Moon is so much closer than the planets, the angular separation will vary depending upon your latitude upon the Earth. With that in mind, I've stuck to stating that the Moon simply appears *close* to a particular planet, star or star cluster in the sky.

	Standard Time	Daylight Savings Time
Greenwich Mean Time	UT-0 hours	UT+1 hours
Eastern Time	UT-5 hours	UT-4 hours
Central Time	UT-6 hours	UT-5 hours
Mountain Time	UT-7 hours	UT-6 hours
Pacific Time	UT-8 hours	UT-7 hours

Complete List of Constellations

Latin	Genitive	English Name	Abbreviation	Size
Andromeda	Andromedae	The Princess	And	19th
Antila	Antilae	The Air Pump	Ant	62nd
Apus	Apodis	The Bird of Paradise	Aps	67th
Aquarius	Aquarii	The Water Bearer	Aqr	10th
Aquila	Aquilae	The Eagle	Aql	22nd
Ara	Arae	The Altar	Ara	63rd
Aries	Arietis	The Ram	Ari	39th
Auriga	Aurigae	The Charioteer	Aur	21st
Boötes	Boötis	The Herdsman	Boo	13th
Caelum	Caeli	The Chisel	Cae	81st
Camelopardalis	Camelopardalis	The Giraffe	Cam	18th
Cancer	Cancri	The Crab	Cnc	31st
Canes Venatici	Canum Venaticorum	The Hunting Dogs	CVn	38th
Canis Major	Canis Majoris	The Large Dog	CMa	43rd
Canis Minor	Canis Minoris	The Little Dog	CMi	71st
Capricornus	Capricorni	The Sea Goat	Cap	40th
Carina	Carinae	The Keel	Car	34th
Cassiopeia	Cassiopeiae	The Queen	Cas	25th
Centaurus	Centauri	The Centaur	Cen	9th
Cepheus	Cephei	The King	Cep	27th
Cetus	Ceti	The Sea Monster	Cet	4th
Chamaeleon	Chamaeleontis	The Chameleon	Cha	79th
Circinus	Circini	The Compass	Cir	85th
Columba	Columbae	The Dove	Col	54th
Coma Berenices	Comae Berenices	Berenices' Hair	Com	42nd
Corona Australis	Coronae Australis	The Southern Crown	CrA	80th
Corona Borealis	Coronae Borealis	The Northern Crown	CrB	73rd
Corvus	Corvi	The Crow	Crv	70th
Crater	Crateris	The Cup	Crt	53rd
Crux	Crucis	The Southern Cross	Cru	88th
Cygnus	Cygni	The Swan	Cyg	16th
Delphinus	Delphini	The Dolphin	Del	69th
Dorado	Doradus	The Goldfish	Dor	72nd
Draco	Draconis	The Dragon	Dra	8th
Equuleus	Equulei	The Foal	Equ	87th
Eridanus	Eridani	The River	Eri	6th

Latin	Genitive	English Name	Abbreviation	Size
Fornax	Fornacis	The Furnace	For	41st
Gemini	Geminorum	The Twins	Gem	30th
Grus	Gruis	The Crane	Gru	45th
Hercules	Herculis	The Hero	Her	5th
Horologium	Horologii	The Clock	Hor	58th
Hydra	Hydrae	The Water Snake	Hya	1st
Hydrus	Hydri	The Lesser Water Snake	Hyi	61st
Indus	Indi	The Indian	Ind	49th
Lacerta	Lacertae	The Lizard	Lac	68th
Leo	Leonis	The Lion	Leo	12th
Leo Minor	Leonis Minoris	The Little Lion	LMi	64th
Lepus	Leporis	The Hare	Lep	51st
Libra	Librae	The Scales	Lib	29th
Lupus	Lupi	The Wolf	Lup	46th
Lynx	Lyncis	The Lynx	Lyn	28th
Lyra	Lyrae	The Lyre	Lyr	52nd
Mensa	Mensae	The Table Mountain	Men	75th
Microscopium	Microscopii	The Microscope	Mic	66th
Monoceros	Monocerotis	The Unicorn	Mon	35th
Musca	Muscae	The Fly	Mus	77th
Norma	Normae	The Carpenter's Level	Nor	74th
Octans	Octantis	The Octant	Oct	50th
Ophiuchus	Ophiuchi	The Serpent Bearer	Oph	11th
Orion	Orionis	The Hunter	Ori	26th
Pavo	Pavonis	The Peacock	Pav	44th
Pegasus	Pegasi	The Flying Horse	Peg	7th
Perseus	Persei	The Hero	Per	24th
Phoenix	Phoenicis	The Phoenix	Phe	37th
Pictor	Pictoris	The Painter's Easel	Pic	59th
Pisces	Piscium	The Fishes	Psc	14th
Piscis Austrinus	Piscis Austrini	The Southern Fish	PsA	60th
Puppis	Puppis	The Poop Deck	Pup	20th
Pyxis	Pyxidis	The Compass	Pyx	65th
Reticulum	Reticuli	The Net	Ret	82nd

Complete List of Constellations (cont.)

Latin	Genitive	English Name	Abbreviation	Size
Sagitta	Sagittae	The Arrow	Sge	86th
Sagittarius	Sagittarii	The Archer	Sgr	15th
Scorpius	Scorpii	The Scorpion	Sco	33rd
Sculptor	Sculptoris	The Sculptor	Scl	36th
Scutum	Scuti	The Shield	Sct	84th
Serpens	Serpentis	The Serpent	Ser	23rd
Sextans	Sextantis	The Sextant	Sex	47th
Taurus	Tauri	The Bull	Tau	17th
Telescopium	Telescopii	The Telescope	Tel	57th
Triangulum	Trianguli	The Triangle	Tri	78th
Triangulum Australe	Trianguli Australis	The Southern Triangle	TrA	83rd
Tucana	Tucanae	The Toucan	Tuc	48th
Ursa Major	Ursae Majoris	The Great Bear	UMa	3rd
Ursa Minor	Ursae Minoris	The Little Bear	UMi	56th
Vela	Velorum	The Sails	Vel	32nd
Virgo	Virginis	The Virgin	Vir	2nd
Volans	Volantis	The Flying Fish	Vol	76th
Vulpecula	Vulpeculae	The Fox	Vul	55th

January 1st to 10th

The Moon

| 1st | 3rd | 5th | 7th | 9th |

Date	Con	R.A.	Dec	Mag	Diam	Ill.	Elon.	Phase	Close To
1st	Aqr	23h 38m	-7° 48'	-9.4	30'	34%	73° E	+Cr	Neptune
2nd	Psc	0h 22m	-3° 16'	-9.8	30'	43%	83° E	FQ	
3rd	Cet	1h 6m	1° 24'	-10.2	30'	53%	93° E	FQ	
4th	Psc	1h 50m	6° 3'	-10.6	30'	62%	103° E	FQ	Uranus
5th	Cet	2h 36m	10° 33'	-11.0	30'	71%	113° E	+G	Uranus
6th	Ari	3h 24m	14° 43'	-11.3	30'	80%	124° E	+G	Pleiades
7th	Tau	4h 15m	18° 20'	-11.6	31'	87%	136° E	+G	Pleiades, Hyades, Aldebaran
8th	Tau	5h 10m	21° 7'	-12.0	31'	94%	148° E	+G	Hyades, Aldebaran
9th	Gem	6h 8m	22° 49'	-12.3	32'	98%	162° E	FM	
10th	Gem	7h 8m	23° 11'	-12.6	32'	100%	176° E	FM	

Mercury and Venus

Mercury
5th

Venus
5th

Mercury

Date	Con.	R.A.	Dec.	Mag.	Diam.	Ill.	Elon.	Vis.	Rat.	Close To
1st	Sgr	18h 22m	-24° 39'	-0.9	5"	99%	6° W	NV	N/A	Jupiter
3rd	Sgr	18h 36m	-24° 39'	-1.0	5"	99%	4° W	NV	N/A	Jupiter
5th	Sgr	18h 50m	-24° 33'	-1.0	5"	100%	3° W	NV	N/A	Jupiter
7th	Sgr	19h 4m	-24° 22'	-1.1	5"	100%	2° W	NV	N/A	Jupiter, Saturn
9th	Sgr	19h 18m	-24° 5'	-1.2	5"	100%	0° W	NV	N/A	Saturn

Venus

Date	Con.	R.A.	Dec.	Mag.	Diam.	Ill.	Elon.	Vis.	Rat.	Close To
1st	Cap	21h 12m	-18° 5'	-4.0	13"	82%	37° E	PM	**	
3rd	Cap	21h 21m	-17° 19'	-4.0	13"	82%	37° E	PM	**	
5th	Cap	21h 31m	-16° 32'	-4.0	13"	81%	37° E	PM	**	
7th	Cap	21h 41m	-15° 43'	-4.0	13"	81%	37° E	PM	**	
9th	Cap	21h 50m	-14° 52'	-4.0	14"	80%	37° E	PM	**	

Mars and the Outer Planets

Mars
5th

Jupiter
5th

Saturn
5th

Mars

Date	Con.	R.A.	Dec.	Mag.	Diam.	Ill.	Elon.	Vis.	Rat.	Close To
1st	Lib	15h 46m	-19° 31'	1.6	4"	96%	45° W	AM	*	
5th	Lib	15h 57m	-20° 8'	1.5	4"	95%	46° W	AM	*	Antares
10th	Sco	16h 11m	-20° 50'	1.5	4"	95%	48° W	AM	*	Antares

The Outer Planets

Planet	Date	Con.	R.A.	Dec.	Mag.	Diam.	Elon.	Vis.	Rat.	Close To
Jupiter	5th	Sgr	18h 33m	-23° 8'	-1.8	32"	7° W	NV	N/A	Mercury
Saturn	5th	Sgr	19h 34m	-21° 36'	0.5	15"	8° E	NV	N/A	
Uranus	5th	Ari	2h 2m	11° 56'	5.7	4"	105° E	PM	***	Moon
Neptune	5th	Aqr	23h 11m	-6° 20'	7.9	2"	62° E	PM	**	

Highlights

Date	Time (UT)	Event
1st	07:10	Asteroid Vesta is stationary prior to resuming prograde motion. (Evening sky.)
3rd	04:46	First Quarter Moon. (Evening sky.)
4th	17:57	The just-past first quarter Moon is south of Uranus. (Evening sky.)
	N/A	The Quadrantid meteor shower is at its maximum. (ZHR: 120)
6th	23:46	The waxing gibbous Moon is south of the Pleiades star cluster. (Taurus, evening sky.)
7th	21:01	The waxing gibbous Moon is north of the bright star Aldebaran. (Taurus, evening sky.)
10th	15:02	Mercury is at superior conjunction with the Sun. (Not visible.)
	19:09	Penumbral lunar eclipse. Visible from Africa, Asia, the Atlantic, Australia, Europe, the Indian Ocean and the far north-eastern and northern North America.
	19:22	Full Moon. (Visible all night.)

Planet Locations – January 5th

Sun · Mercury · Venus · Mars · Jupiter · Saturn

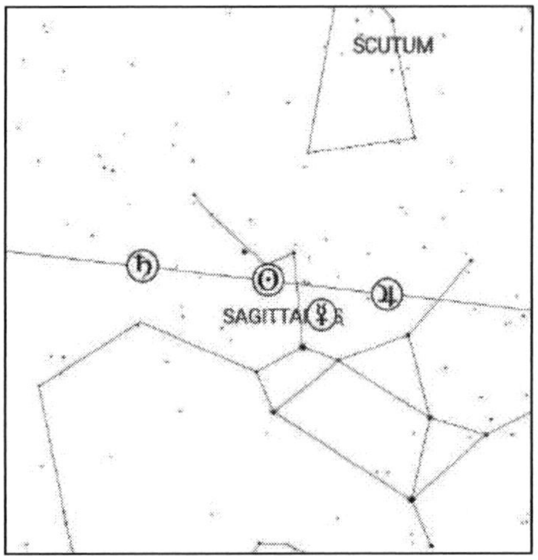

Sun

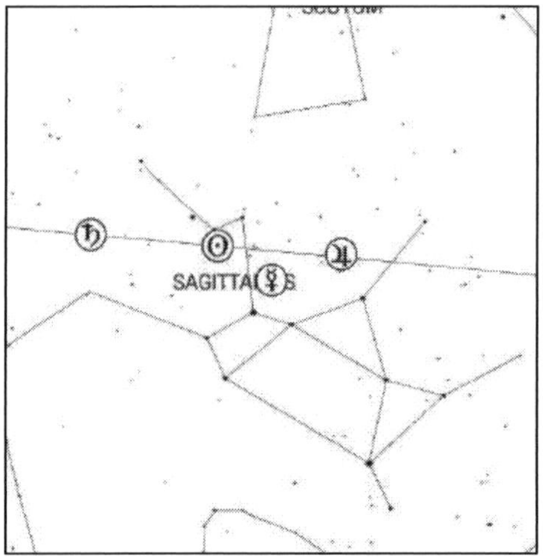

Mercury

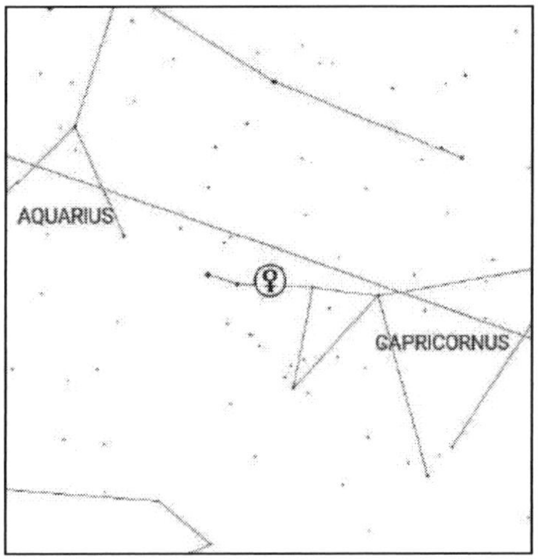

Venus

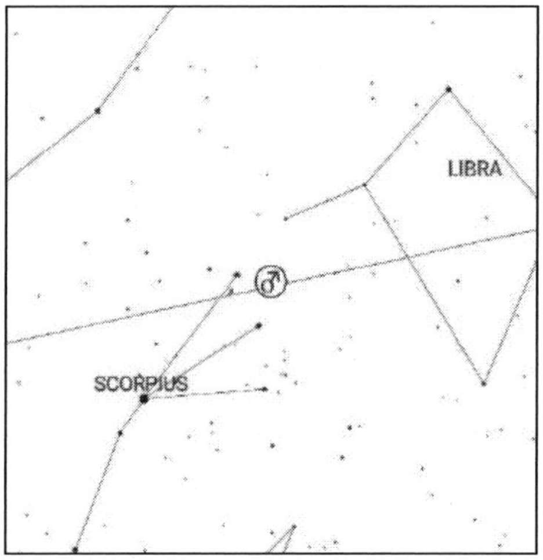

Mars

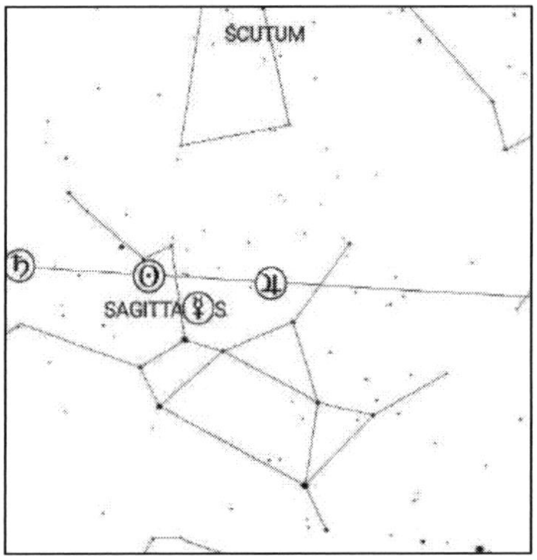

Jupiter

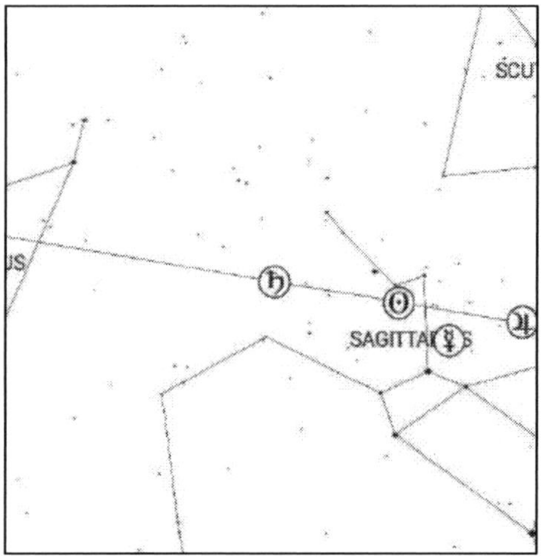

Saturn

January 3rd to 4th – Evening & Morning Sky

The Quadrantid meteor shower is at its best from late on January 3rd through to the early hours of January 4th.

This year, the light from the first quarter Moon might interfere on the evening of the 3rd, making it difficult to see the fainter meteors. If you're able to stay up late (or get up early), you'll have a better chance in the early hours of the 4th, after the Moon has set.

This image depicts the sky looking north-east at 2 a.m.

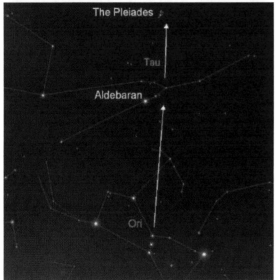

January 5th – Evening Sky

If you're new to astronomy, you can use the familiar constellation of Orion (Ori), the Hunter, to locate objects of interest in the sky.

For example, by following the three stars of his belt upwards, you arrive at Aldebaran, the brightest star in the constellation of Taurus (Tau) the Bull. Continue on to find the Pleiades star cluster – a stunning sight in binoculars!

This image depicts the sky looking south-east at about 8p.m.

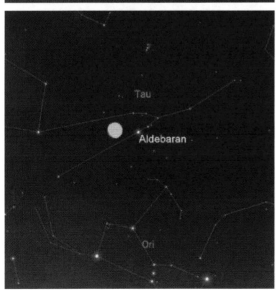

January 7th – Evening Sky

The Moon turns full in three days' time but tonight it appears close to Aldebaran.

The brightest star in the constellation of Taurus (Tau), the Bull, Aldebaran represents the red eye of the beast. The light from the waxing gibbous Moon might make fainter stars harder to see, but you shouldn't have any problem glimpsing Aldebaran beside it.

This image depicts the sky looking south-east, about two hours after sunset.

January 11th to 20th, 2020

The Moon

| 11th | 13th | 15th | 17th | 19th |

Date	Con	R.A.	Dec	Mag	Diam	Ill.	Elon.	Phase	Close To
11th	Cnc	8h 9m	22° 4'	-12.5	32'	99%	170° W	FM	Praesepe
12th	Cnc	9h 10m	19° 31'	-12.1	33'	96%	156° W	FM	Praesepe
13th	Leo	10h 9m	15° 43'	-11.8	33'	90%	142° W	-G	Regulus
14th	Leo	11h 5m	10° 58'	-11.4	33'	82%	129° W	-G	
15th	Vir	12h 0m	5° 38'	-11.0	33'	73%	117° W	-G	
16th	Vir	12h 52m	0° 3'	-10.6	32'	62%	105° W	LQ	Spica
17th	Vir	13h 44m	-5° 29'	-10.1	32'	50%	93° W	LQ	Spica
18th	Lib	14h 37m	-10° 39'	-9.6	32'	39%	81° W	LQ	
19th	Lib	15h 30m	-15° 12'	-9.0	32'	29%	68° W	-Cr	
20th	Oph	16h 24m	-18° 53'	-8.4	31'	20%	56° W	-Cr	Mars, Antares

Mercury and Venus

Mercury
15th

Venus
15th

Mercury

Date	Con.	R.A.	Dec.	Mag.	Diam.	Ill.	Elon.	Vis.	Rat.	Close To
11th	Sgr	19h 33m	-23° 42'	-1.2	5"	100%	1° E	NV	N/A	Saturn
13th	Sgr	19h 47m	-23° 13'	-1.2	5"	100%	2° E	NV	N/A	Saturn
15th	Sgr	20h 1m	-22° 37'	-1.2	5"	99%	4° E	NV	N/A	Saturn
17th	Cap	20h 15m	-21° 56'	-1.2	5"	99%	5° E	NV	N/A	Saturn
19th	Cap	20h 29m	-21° 8'	-1.1	5"	98%	7° E	NV	N/A	

Venus

Date	Con.	R.A.	Dec.	Mag.	Diam.	Ill.	Elon.	Vis.	Rat.	Close To
11th	Aqr	22h 0m	-13° 60'	-4.0	14"	79%	38° E	PM	**	
13th	Aqr	22h 9m	-13° 6'	-4.0	14"	79%	38° E	PM	**	
15th	Aqr	22h 18m	-12° 11'	-4.0	14"	78%	38° E	PM	**	
17th	Aqr	22h 27m	-11° 15'	-4.0	14"	78%	38° E	PM	**	
19th	Aqr	22h 36m	-10° 18'	-4.1	14"	77%	38° E	PM	**	

Mars and the Outer Planets

Mars
15th

Jupiter
15th

Saturn
15th

Mars

Date	Con.	R.A.	Dec.	Mag.	Diam.	Ill.	Elon.	Vis.	Rat.	Close To
11th	Sco	16h 14m	-20° 58'	1.5	4"	95%	49° W	AM	*	Antares
15th	Oph	16h 25m	-21° 28'	1.5	5"	95%	50° W	AM	*	Antares
20th	Oph	16h 40m	-22° 1'	1.4	5"	94%	52° W	AM	*	Moon, Antares

The Outer Planets

Planet	Date	Con.	R.A.	Dec.	Mag.	Diam.	Elon.	Vis.	Rat.	Close To
Jupiter	15th	Sgr	18h 43m	-22° 60'	-1.8	32"	16° W	AM	*	
Saturn	15th	Sgr	19h 39m	-21° 25'	0.5	15"	2° W	NV	N/A	Mercury
Uranus	15th	Ari	2h 2m	11° 56'	5.8	4"	94° E	PM	**	
Neptune	15th	Aqr	23h 12m	-6° 14'	7.9	2"	51° E	PM	**	

Highlights

Date	Time (UT)	Event
11th	05:38	Uranus is stationary prior to resuming prograde motion. (Evening sky.)
	23:15	The just-past full Moon is north of the Praesepe open star cluster. (Cancer, morning sky.)
13th	09:26	Dwarf planet Pluto is in conjunction with the Sun. (Not visible.)
	12:17	The waning gibbous Moon is north of the bright star Regulus. (Leo, morning sky.)
	15:09	Saturn is in conjunction with the Sun. (Not visible.)
14th	09:58	Dwarf planet Ceres is in conjunction with the Sun. (Not visible.)
16th	17:45	Mars is 4.8° north of the bright star Antares. (Scorpius, morning sky.)
17th	01:51	The almost last quarter Moon is north of the bright star Spica. (Virgo, morning sky.)
	12:59	Last Quarter Moon. (Morning sky.)
20th	15:24	The waning crescent Moon is north of the bright star Antares. (Scorpius, morning sky.)
	19:39	The waning crescent Moon is north of Mars. (Morning sky.)

Planet Locations – January 15th

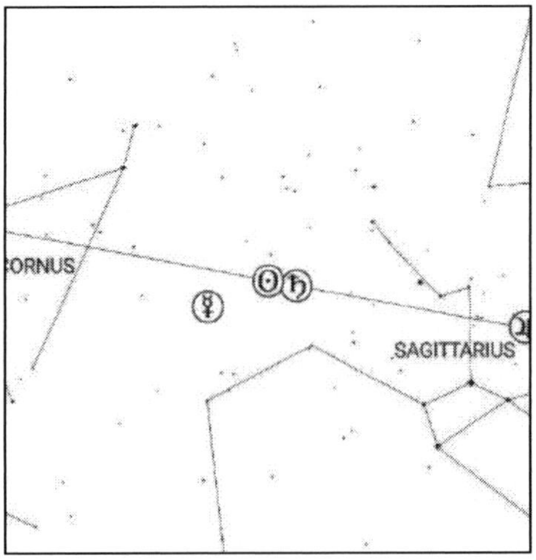

Sun

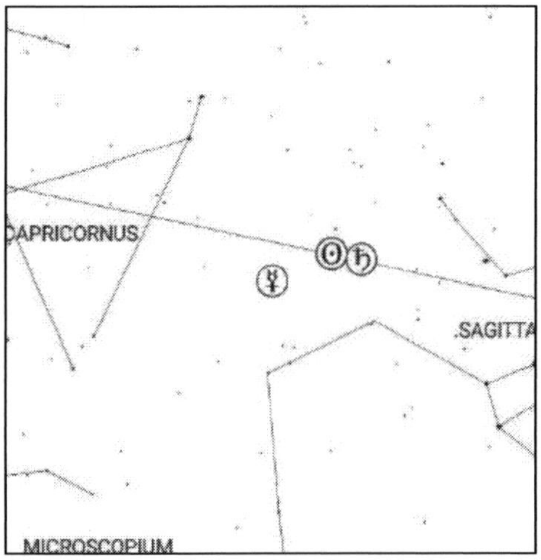

Mercury

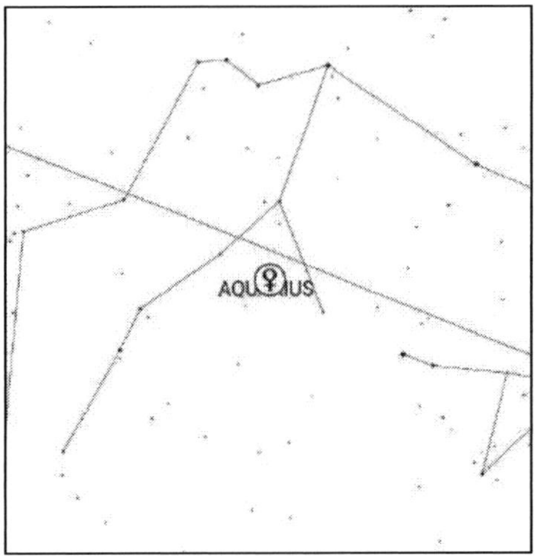

Venus

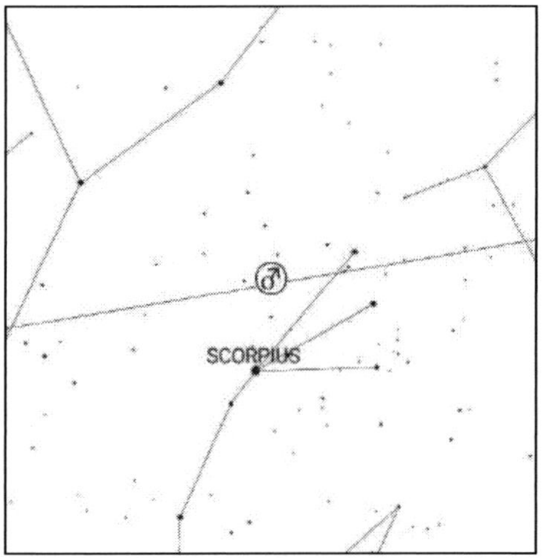

Mars

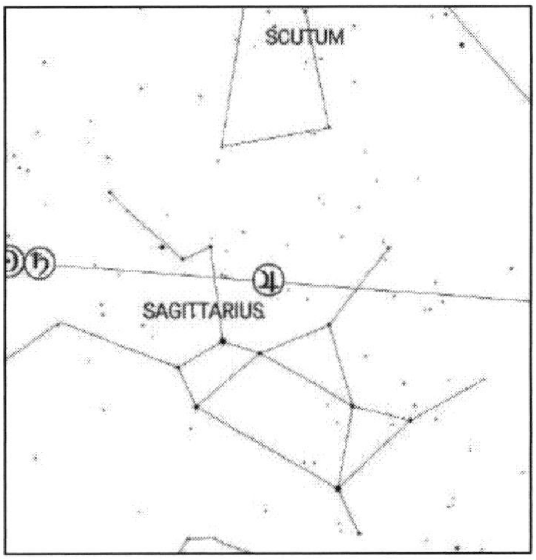

Jupiter

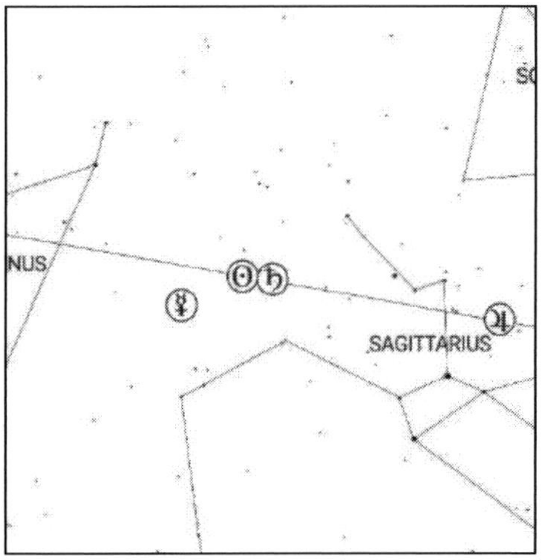

Saturn

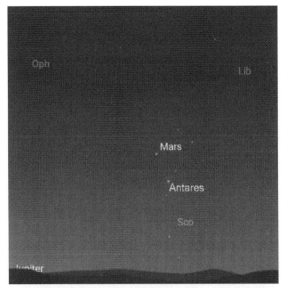

January 16th – Morning Sky

The planet Mars is slowly passing the bright star Antares in the predawn sky. This morning the pair are at their closest, with about five degrees between them. The name Antares literally means "rival of Mars" because the star appears to be of a similar color to our nearby planetary neighbor.

Take a moment to compare the two this morning. Which appears brighter and what are their respective colors?

This image depicts the sky looking south-east at about 45 minutes before sunrise.

January 17th – Morning Sky

The Moon reaches last quarter this morning and can be found close to Spica, the brightest star in the constellation of Virgo (Vir), the Virgin.

A spring constellation, it's home to a number of galaxies that are a favorite target for telescopic observers.

Also look out for the orange giant star Arcturus, high above the horizon in the pre-dawn sky.

This image depicts the sky looking south at about 90 minutes before sunrise.

January 20th – Morning Sky

The Moon has now moved ahead to Scorpius (Sco), the Scorpion, and can be found close to Mars in the pre-dawn sky.

Look for the red planet just to the lower left of the crescent Moon in the morning twilight.

Do you have an unobstructed view? Can you see Jupiter just beginning to rise above the horizon? If not, come back in a few days and try your luck again.

This image depicts the sky looking south-east at about 45 minutes before sunrise.

January 21st to 31st, 2020

The Moon

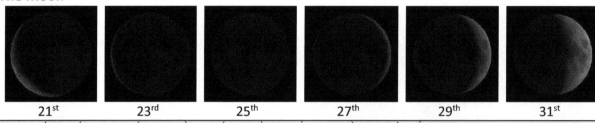

| 21st | 23rd | 25th | 27th | 29th | 31st |

Date	Con	R.A.	Dec	Mag	Diam	Ill.	Elon.	Phase	Close To
21st	Oph	17h 20m	-21° 31'	-7.6	31'	12%	43° W	NM	Mars
22nd	Sgr	18h 16m	-22° 58'	-6.7	31'	6%	30° W	NM	Jupiter
23rd	Sgr	19h 12m	-23° 9'	-5.6	31'	2%	17° W	NM	Jupiter, Saturn
24th	Sgr	20h 7m	-22° 8'	-4.4	30'	0%	4° W	NM	Saturn
25th	Cap	21h 0m	-20° 2'	-4.7	30'	0%	8° E	NM	Mercury
26th	Cap	21h 50m	-17° 2'	-5.8	30'	3%	19° E	NM	Mercury
27th	Aqr	22h 38m	-13° 21'	-6.8	30'	6%	30° E	NM	Venus, Neptune
28th	Aqr	23h 24m	-9° 10'	-7.6	30'	12%	41° E	NM	Venus, Neptune
29th	Psc	0h 8m	-4° 42'	-8.3	29'	19%	51° E	+Cr	Venus
30th	Cet	0h 51m	0° 4'	-8.9	29'	27%	61° E	+Cr	
31st	Psc	1h 35m	4° 35'	-9.4	30'	35%	70° E	+Cr	Uranus

Mercury and Venus

Mercury
25th

Venus
25th

Mercury

Date	Con.	R.A.	Dec.	Mag.	Diam.	Ill.	Elon.	Vis.	Rat.	Close To
21st	Cap	20h 44m	-20° 14'	-1.1	5"	97%	8° E	NV	N/A	
23rd	Cap	20h 58m	-19° 14'	-1.1	5"	96%	9° E	NV	N/A	
25th	Cap	21h 11m	-18° 9'	-1.1	5"	94%	11° E	NV	N/A	Moon
27th	Cap	21h 25m	-16° 57'	-1.1	5"	92%	12° E	NV	N/A	
29th	Cap	21h 38m	-15° 41'	-1.0	5"	89%	13° E	NV	N/A	
31st	Cap	21h 51m	-14° 20'	-1.0	6"	86%	14° E	NV	N/A	

Venus

Date	Con.	R.A.	Dec.	Mag.	Diam.	Ill.	Elon.	Vis.	Rat.	Close To
21st	Aqr	22h 45m	-9° 19'	-4.1	14"	77%	38° E	PM	**	
23rd	Aqr	22h 54m	-8° 20'	-4.1	15"	76%	38° E	PM	**	Neptune
25th	Aqr	23h 3m	-7° 20'	-4.1	15"	75%	39° E	PM	**	Neptune
27th	Aqr	23h 12m	-6° 19'	-4.1	15"	75%	39° E	PM	**	Moon, Neptune
29th	Aqr	23h 20m	-5° 18'	-4.1	15"	74%	39° E	PM	**	Moon, Neptune
31st	Aqr	23h 29m	-4° 16'	-4.1	15"	74%	39° E	PM	**	Neptune

Mars and the Outer Planets

Mars
25th

Jupiter
25th

Saturn
25th

Mars

Date	Con.	R.A.	Dec.	Mag.	Diam.	Ill.	Elon.	Vis.	Rat.	Close To
21st	Oph	16h 43m	-22° 7'	1.4	5"	94%	52° W	AM	*	Moon, Antares
25th	Oph	16h 54m	-22° 30'	1.4	5"	94%	54° W	AM	*	Antares
31st	Oph	17h 12m	-22° 58'	1.4	5"	93%	55° W	AM	*	

The Outer Planets

Planet	Date	Con.	R.A.	Dec.	Mag.	Diam.	Elon.	Vis.	Rat.	Close To
Jupiter	25th	Sgr	18h 53m	-22° 49'	-1.9	32"	24° W	AM	*	
Saturn	25th	Sgr	19h 44m	-21° 14'	0.6	15"	11° W	NV	N/A	
Uranus	25th	Ari	2h 2m	11° 58'	5.8	4"	83° E	PM	**	
Neptune	25th	Aqr	23h 13m	-6° 7'	7.9	2"	41° E	PM	**	Venus

Highlights

Date	Time (UT)	Event
21st	N/A	Good opportunity to see Earthshine on the waning crescent Moon. (Morning sky.)
23rd	01:50	The waning crescent Moon is south of Jupiter. (Morning sky.)
24th	21:43	New Moon. (Not visible.)
27th	19:25	Venus is 0.1° south of Neptune. (Evening sky.)
28th	05:29	The waning crescent Moon is south of Neptune. (Evening sky.)
	06:19	The waning crescent Moon is south of Venus. (Evening sky.)
	N/A	Good opportunity to see Earthshine on the waxing crescent Moon. (Evening sky.)

Planet Locations – January 25th

Sun	Moon	Mercury	Venus	Mars	Jupiter	Saturn

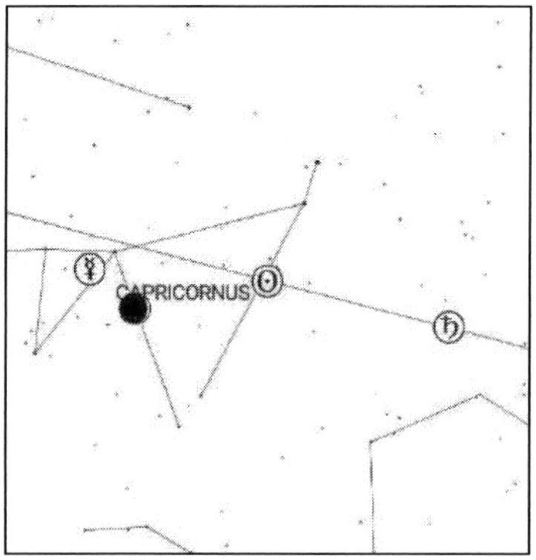

Sun

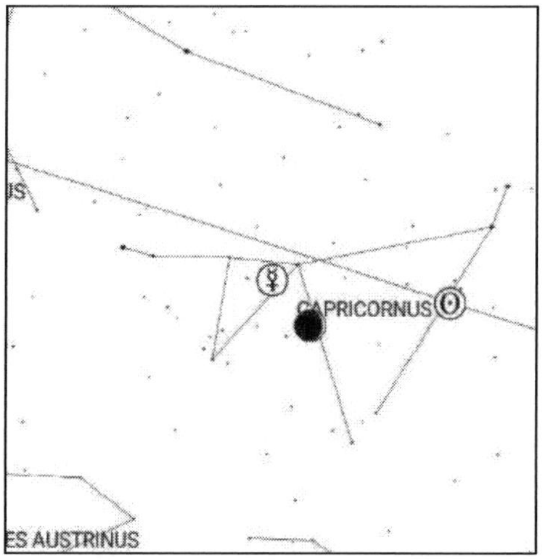

Mercury

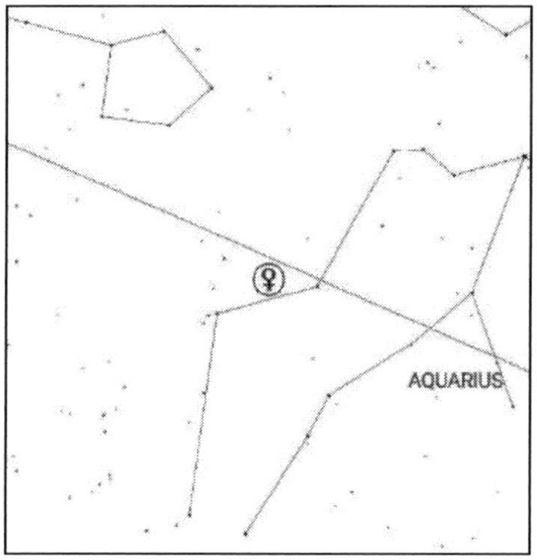

Venus

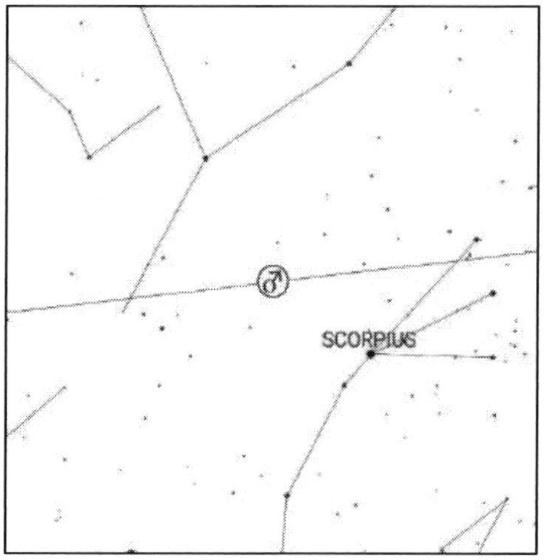

Mars

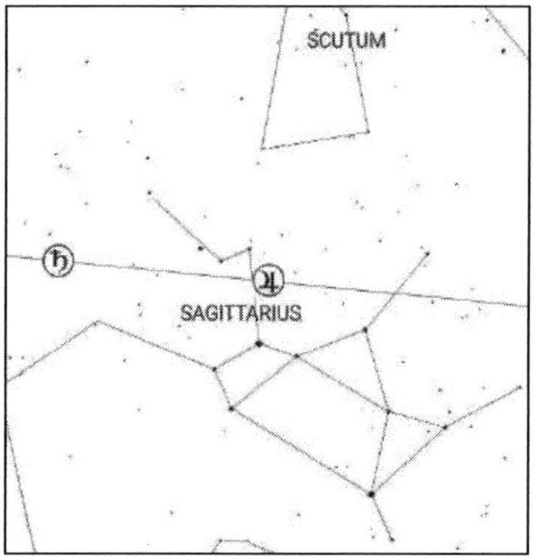

Jupiter

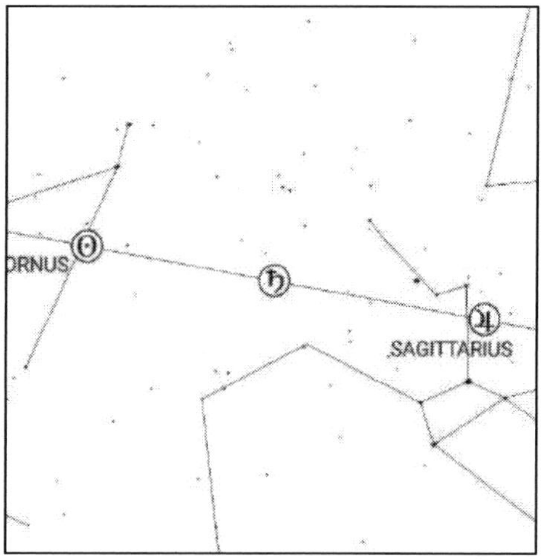

Saturn

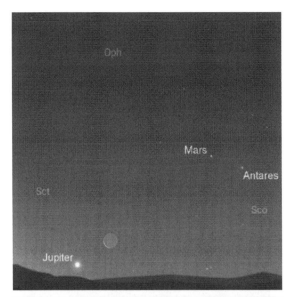

January 22nd – Morning Sky

If you had difficulty find Jupiter a few days ago, try again this morning.

You'll need a clear, unobstructed view of the horizon, but if you can see the thin, crescent Moon, you may be able to glimpse our solar system's largest planet to the lower left of it.

This image depicts the sky looking south-east at about 45 minutes before sunrise.

January 27th – Evening Sky

The Moon turned new three days ago and has returned to the early evening sky. It can be found below brilliant Venus in the twilight tonight – but can you also spot Mercury clinging to the horizon?

The tiny planet is slowly creeping away from the Sun and will be better seen in a few weeks time.

This image depicts the sky looking south-west at about 30 minutes after sunset.

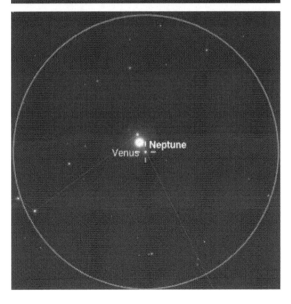

January 27th – Evening Sky

Here's something you don't see every day – or night – a close conjunction of Venus and Neptune. Although Venus is very easily seen with just the unaided eye, you'll need binoculars to see faint Neptune. You can find it just to the lower right of its brighter sibling.

If it's cloudy, come back tomorrow. Although the pair are closest today, they'll continue to be close (but further apart) until the end of the month.

This image depicts the view through 10x50 binoculars, looking west about two hours after sunset.

February 1st to 10th, 2020

The Moon

1st

3rd

5th

7th

9th

Date	Con	R.A.	Dec	Mag	Diam	Ill.	Elon.	Phase	Close To
1st	Cet	2h 20m	9° 6'	-9.9	30'	45%	80° E	FQ	Uranus
2nd	Ari	3h 6m	13° 20'	-10.3	30'	54%	91° E	FQ	Pleiades
3rd	Tau	3h 55m	17° 6'	-10.7	30'	64%	102° E	FQ	Pleiades, Hyades, Aldebaran
4th	Tau	4h 47m	20° 11'	-11.1	31'	73%	114° E	+G	Hyades, Aldebaran
5th	Tau	5h 43m	22° 19'	-11.4	31'	82%	127° E	+G	
6th	Gem	6h 41m	23° 15'	-11.7	32'	90%	141° E	+G	
7th	Gem	7h 42m	22° 46'	-12.1	32'	95%	155° E	FM	
8th	Cnc	8h 44m	20° 46'	-12.4	33'	99%	170° E	FM	Praesepe
9th	Leo	9h 45m	17° 22'	-12.6	33'	100%	176° W	FM	Regulus
10th	Leo	10h 44m	12° 48'	-12.3	33'	98%	162° W	FM	Regulus

Mercury and Venus

Mercury
5th

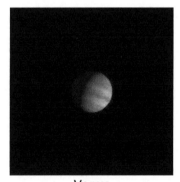
Venus
5th

Mercury

Date	Con.	R.A.	Dec.	Mag.	Diam.	Ill.	Elon.	Vis.	Rat.	Close To
1st	Cap	21h 57m	-13° 38'	-1.0	6"	84%	15° E	NV	N/A	
3rd	Aqr	22h 9m	-12° 13'	-1.0	6"	79%	16° E	PM	**	
5th	Aqr	22h 20m	-10° 47'	-0.9	6"	72%	17° E	PM	**	
7th	Aqr	22h 30m	-9° 22'	-0.8	7"	65%	17° E	PM	**	
9th	Aqr	22h 39m	-8° 1'	-0.6	7"	57%	17° E	PM	***	

Venus

Date	Con.	R.A.	Dec.	Mag.	Diam.	Ill.	Elon.	Vis.	Rat.	Close To
1st	Aqr	23h 33m	-3° 45'	-4.1	15"	73%	39° E	PM	**	Neptune
3rd	Psc	23h 42m	-2° 43'	-4.1	16"	73%	39° E	PM	**	
5th	Psc	23h 50m	-1° 40'	-4.1	16"	72%	39° E	PM	**	
7th	Psc	23h 59m	0° 37'	-4.1	16"	71%	39° E	PM	**	
9th	Psc	0h 7m	0° 26'	-4.1	16"	71%	39° E	PM	**	

Mars and the Outer Planets

Mars	Jupiter	Saturn
5th	5th	5th

Mars

Date	Con.	R.A.	Dec.	Mag.	Diam.	Ill.	Elon.	Vis.	Rat.	Close To
1st	Oph	17h 15m	-23° 2'	1.4	5"	93%	56° W	AM	*	
5th	Oph	17h 27m	-23° 16'	1.3	5"	93%	57° W	AM	*	
10th	Oph	17h 42m	-23° 29'	1.3	5"	93%	58° W	AM	*	

The Outer Planets

Planet	Date	Con.	R.A.	Dec.	Mag.	Diam.	Elon.	Vis.	Rat.	Close To
Jupiter	5th	Sgr	19h 3m	-22° 36'	-1.9	33"	33° W	AM	*	
Saturn	5th	Sgr	19h 50m	-21° 1'	0.6	15"	21° W	AM	**	
Uranus	5th	Ari	2h 3m	12° 2'	5.8	4"	72° E	PM	**	
Neptune	5th	Aqr	23h 14m	-5° 59'	8.0	2"	30° E	PM	*	Venus

Highlights

Date	Time (UT)	Event
1st	03:38	The nearly first quarter Moon is south of Uranus. (Evening sky.)
2nd	01:42	First Quarter Moon. (Evening sky.)
3rd	07:47	The just-past first quarter Moon Is south of the Pleiades star cluster. (Taurus, evening sky.)
4th	07:05	The waxing gibbous Moon is north of the bright star Aldebaran. (Taurus, evening sky.)
8th	10:29	The nearly full Moon is north of the Praesepe star cluster. (Cancer, evening sky.)
9th	07:34	Full Moon. (Visible all night.)
	20:05	The full Moon is north of the bright star Regulus. (Leo, visible all night.)
10th	13:50	Mercury is at greatest eastern elongation from the Sun. (Evening sky.)

Planet Locations – February 5th

Sun Mercury Venus Mars Jupiter Saturn

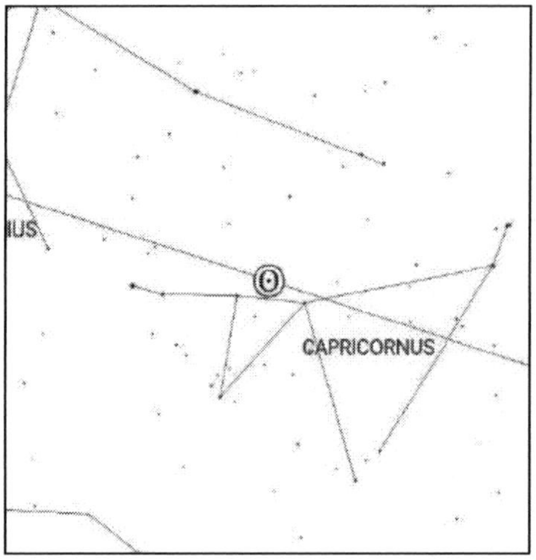

Sun

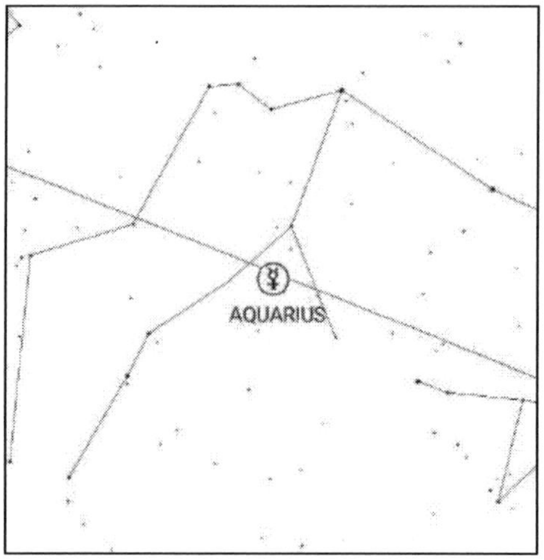

Mercury

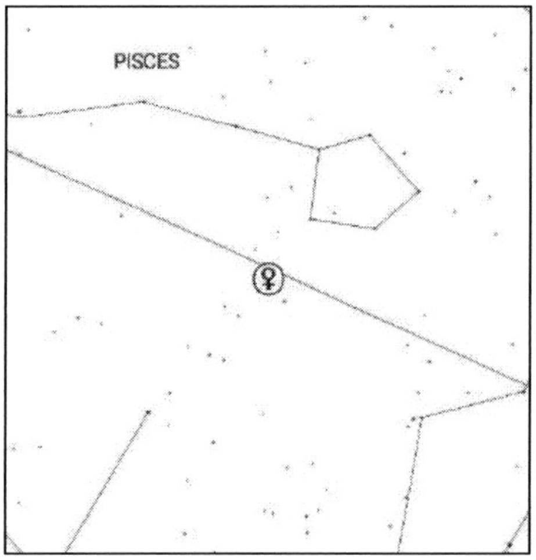

Venus

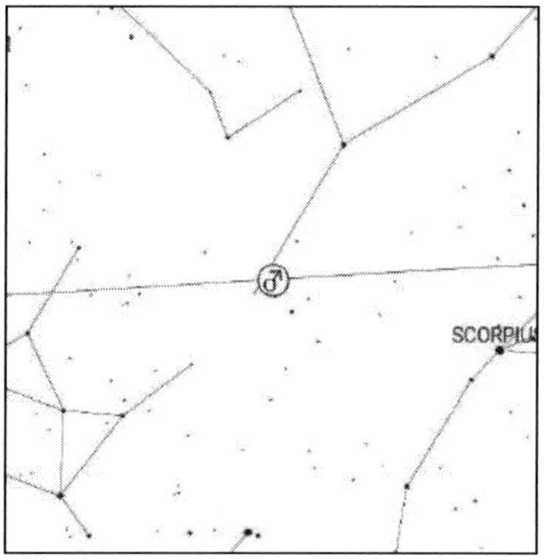

Mars

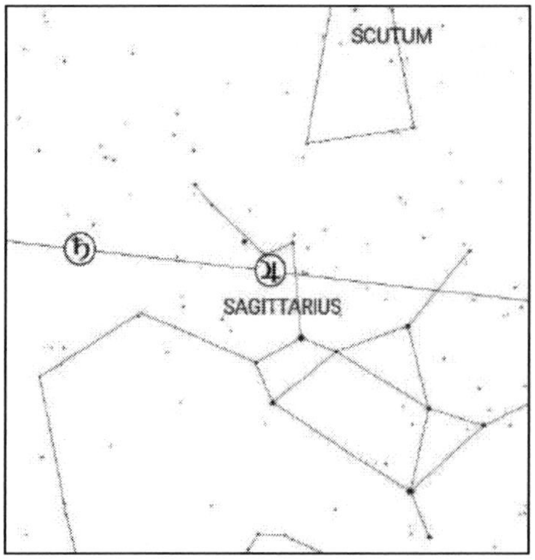

Jupiter

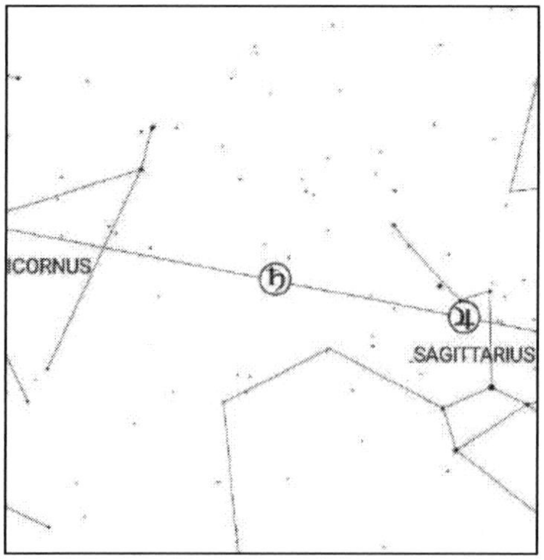

Saturn

February 3rd – Evening Sky

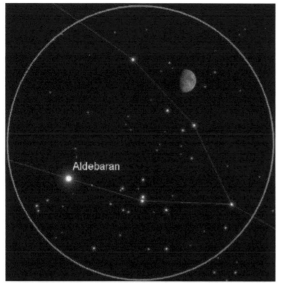

The Moon reached first quarter yesterday and has now returned to the constellation of Taurus (Tau), the Bull.

Tonight it appears close to Aldebaran, the red eye of the Bull. The pair will easily fit within the same binocular field of view, but how many of the surrounding stars can you see?

This image depicts the view through 10x50 binoculars, looking south at about two hours after sunset.

February 9th – Evening Sky

The Moon turned full in the early hours of this morning and now rises at sunset. This evening it appears close to Regulus, the brightest star in the constellation of Leo, the Lion.

It will now slip uneventfully into the pre-dawn sky, but get up early to see an unmissable close encounter between the Moon and Mars on the 18th.

This image depicts the sky looking east at about 9 p.m.

February 10th – Evening Sky

Mercury has now reached greatest elongation from the Sun for the first time this year. This means it's now visible in the evening sky.

Look for it close to the horizon, some way to the lower right of Venus. If you don't spot it, try again tomorrow but it'll probably sink below the horizon within the next four or five days. There'll be better opportunities in June and September.

This image depicts the sky looking west at about 30 minutes after sunset.

February 11ᵗʰ to 20ᵗʰ, 2020

The Moon

    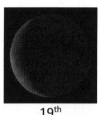

| 11ᵗʰ | 13ᵗʰ | 15ᵗʰ | 17ᵗʰ | 19ᵗʰ |

Date	Con	R.A.	Dec	Mag	Diam	Ill.	Elon.	Phase	Close To
11th	Vir	11h 41m	7° 26'	-11.9	33'	93%	149° W	-G	
12th	Vir	12h 36m	1° 40'	-11.6	33'	86%	136° W	-G	
13th	Vir	13h 30m	-4° 6'	-11.2	33'	76%	124° W	-G	Spica
14th	Lib	14h 23m	-9° 33'	-10.8	32'	66%	112° W	LQ	
15th	Lib	15h 17m	-14° 21'	-10.3	32'	55%	99° W	LQ	
16th	Sco	16h 12m	-18° 17'	-9.8	32'	44%	86° W	LQ	Antares
17th	Oph	17h 7m	-21° 10'	-9.3	31'	33%	74° W	-Cr	Antares
18th	Sgr	18h 3m	-22° 51'	-8.7	31'	24%	61° W	-Cr	Mars
19th	Sgr	18h 58m	-23° 18'	-8.0	31'	16%	48° W	-Cr	Jupiter
20th	Sgr	19h 53m	-22° 34'	-7.2	30'	9%	35° W	NM	Jupiter, Saturn

Mercury and Venus

 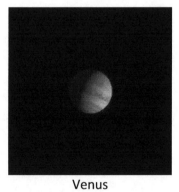

Mercury
15ᵗʰ

Venus
15ᵗʰ

Mercury

Date	Con.	R.A.	Dec.	Mag.	Diam.	Ill.	Elon.	Vis.	Rat.	Close To
11th	Aqr	22h 45m	-6° 48'	-0.4	7"	48%	17° E	PM	***	
13th	Aqr	22h 50m	-5° 44'	0.0	8"	38%	16° E	PM	**	
15th	Aqr	22h 53m	-4° 55'	0.4	8"	29%	15° E	NV	N/A	
17th	Aqr	22h 52m	-4° 22'	1.0	9"	20%	13° E	NV	N/A	
19th	Aqr	22h 50m	-4° 8'	1.8	9"	12%	10° E	NV	N/A	

Venus

Date	Con.	R.A.	Dec.	Mag.	Diam.	Ill.	Elon.	Vis.	Rat.	Close To
11th	Psc	0h 16m	1° 29'	-4.1	16"	70%	39° E	PM	**	
13th	Psc	0h 24m	2° 31'	-4.2	17"	69%	40° E	PM	***	
15th	Psc	0h 32m	3° 34'	-4.2	17"	68%	40° E	PM	***	
17th	Psc	0h 41m	4° 36'	-4.2	17"	68%	40° E	PM	***	
19th	Psc	0h 49m	5° 38'	-4.2	17"	67%	40° E	PM	***	

Mars and the Outer Planets

Mars
15th

Jupiter
15th

Saturn
15th

Mars

Date	Con.	R.A.	Dec.	Mag.	Diam.	Ill.	Elon.	Vis.	Rat.	Close To
11th	Sgr	17h 45m	-23° 31'	1.3	5"	92%	58° W	AM	*	
15th	Sgr	17h 57m	-23° 37'	1.2	5"	92%	59° W	AM	*	
20th	Sgr	18h 12m	-23° 40'	1.2	5"	92%	60° W	AM	*	

The Outer Planets

Planet	Date	Con.	R.A.	Dec.	Mag.	Diam.	Elon.	Vis.	Rat.	Close To
Jupiter	15th	Sgr	19h 12m	-22° 22'	-1.9	33"	40° W	AM	*	
Saturn	15th	Sgr	19h 54m	-20° 49'	0.6	15"	30° W	AM	**	
Uranus	15th	Ari	2h 4m	12° 8'	5.8	3"	63° E	PM	**	
Neptune	15th	Aqr	23h 15m	-5° 51'	8.0	2"	20° E	PM	*	Mercury

Highlights

Date	Time (UT)	Event
13th	10:59	The waning gibbous Moon is north of the bright star Spica. (Virgo, morning sky.)
15th	22:18	Last Quarter Moon. (Morning sky.)
16th	10:04	Mercury is stationary prior to beginning retrograde motion. (Evening sky.)
	19:21	The just-past last quarter Moon is north of the bright star Antares. (Scorpius, morning sky.)
18th	14:39	The waning crescent Moon is north of Mars. (Morning sky.)
19th	20:04	The waning crescent Moon is south of Jupiter. (Morning sky.)
20th	14:57	The waning crescent Moon is south of Saturn. (Morning sky.)
	N/A	Good opportunity to see Earthshine on the waning crescent Moon. (Morning sky.)

Planet Locations – February 15ᵗʰ

☉	☿	♀	♂	♃	♄
Sun	Mercury	Venus	Mars	Jupiter	Saturn

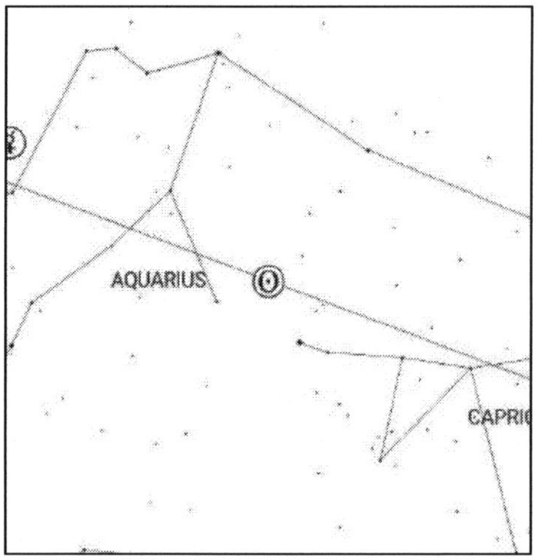

Sun

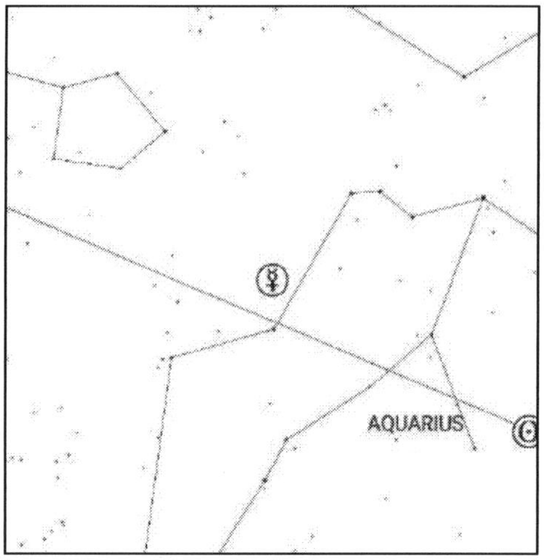

Mercury

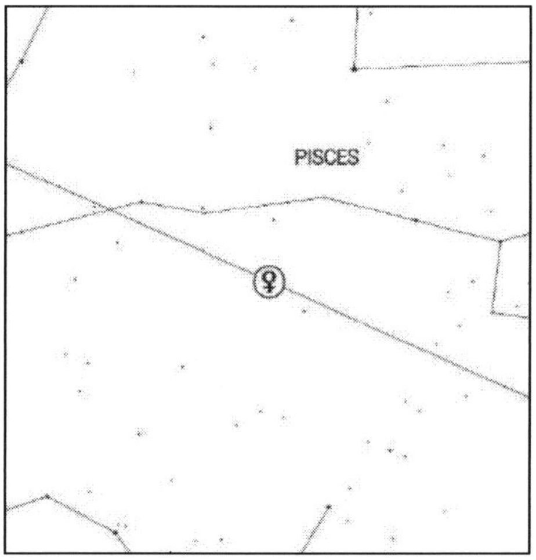

Venus

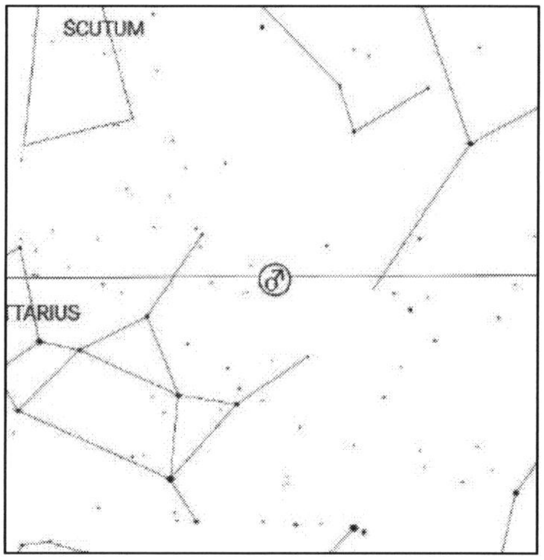

Mars

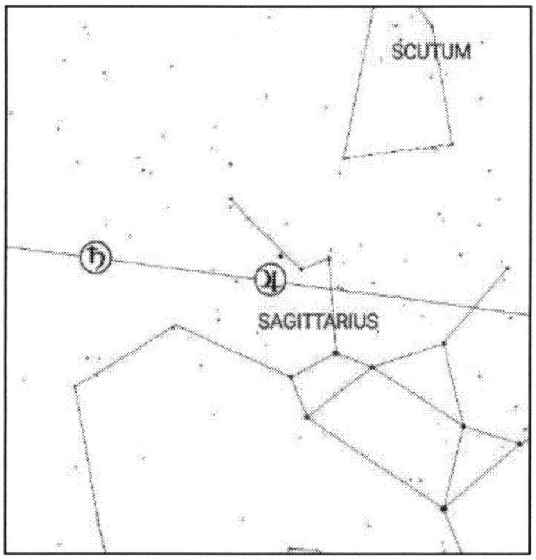

Jupiter

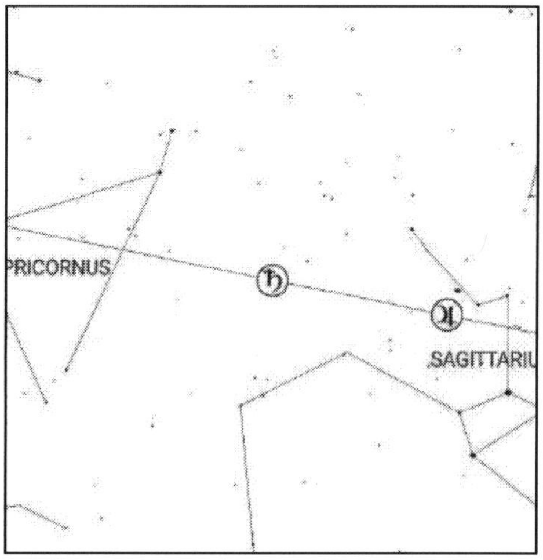

Saturn

February 18th – Morning Sky

The western half of the United States is in for a treat this morning as early rises will have a chance to witness an occultation of Mars by the Moon. A relatively rare event, this occurs when the Moon appears to pass in front of the planet, hiding it from view.

The timing of the event will vary, depending upon your location within the U.S., so it's best to check online for the most accurate information.

This image depicts the view through 10x50 binoculars from Kansas City, KS, at 5:30 a.m. CT

February 19th – Morning Sky

This morning the waning crescent Moon has moved past Mars and now appears to the right of the giant planet Jupiter.

Meanwhile, Saturn hovers to the lower left, a little closer to the horizon. The three planets appear in a straight line, with bright Jupiter nearly midway between the two smaller worlds.

This image depicts the sky looking south-east at about 30 minutes before sunrise.

February 20th – Morning Sky

If you had difficulty spotting Saturn yesterday then the Moon will provide a convenient marker this morning. The ringed planet appears just to the upper left of the thinning crescent.

This is also a good opportunity to see Earthshine illuminating the darkened portion of the Moon's surface. It turns new on the 23rd and will then emerge again into the evening sky.

This image depicts the sky looking south-east at about 30 minutes before sunrise.

February 21st to 29th, 2020

The Moon

21st

23rd

25th

27th

29th

Date	Con	R.A.	Dec	Mag	Diam	Ill.	Elon.	Phase	Close To
21st	Cap	20h 46m	-20° 43'	-6.3	30'	4%	23° W	NM	
22nd	Cap	21h 36m	-17° 57'	-5.3	30'	1%	11° W	NM	
23rd	Aqr	22h 24m	-14° 25'	-4.4	30'	0%	0° E	NM	Mercury
24th	Aqr	23h 11m	-10° 21'	-5.0	30'	1%	11° E	NM	Mercury, Neptune
25th	Aqr	23h 55m	-5° 54'	-6.0	29'	3%	21° E	NM	Neptune
26th	Cet	0h 39m	-1° 16'	-6.9	29'	7%	31° E	NM	Venus
27th	Psc	1h 22m	3° 25'	-7.7	29'	13%	41° E	+Cr	Venus, Uranus
28th	Psc	2h 6m	7° 59'	-8.4	30'	20%	51° E	+Cr	Venus, Uranus
29th	Ari	2h 51m	12° 17'	-9.0	30'	28%	61° E	+Cr	Uranus

Mercury and Venus

Mercury
25th

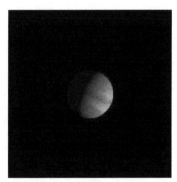
Venus
25th

Mercury

Date	Con.	R.A.	Dec.	Mag.	Diam.	Ill.	Elon.	Vis.	Rat.	Close To
21st	Aqr	22h 45m	-4° 14'	2.7	10"	6%	7° E	NV	N/A	
23rd	Aqr	22h 38m	-4° 38'	3.8	10"	3%	4° E	NV	N/A	Moon
25th	Aqr	22h 31m	-5° 17'	4.6	11"	1%	0° W	NV	N/A	
27th	Aqr	22h 23m	-6° 7'	4.4	11"	1%	4° W	NV	N/A	
29th	Aqr	22h 15m	-7° 3'	3.6	11"	3%	8° W	NV	N/A	

Venus

Date	Con.	R.A.	Dec.	Mag.	Diam.	Ill.	Elon.	Vis.	Rat.	Close To
21st	Psc	0h 57m	6° 39'	-4.2	18"	66%	40° E	PM	***	
23rd	Psc	1h 5m	7° 40'	-4.2	18"	65%	40° E	PM	***	
25th	Psc	1h 14m	8° 40'	-4.2	18"	65%	40° E	PM	***	
27th	Psc	1h 22m	9° 39'	-4.2	18"	64%	41° E	PM	***	Moon
29th	Psc	1h 30m	10° 38'	-4.2	19"	63%	41° E	PM	***	

Mars and the Outer Planets

Mars
25th

Jupiter
25th

Saturn
25th

Mars

Date	Con.	R.A.	Dec.	Mag.	Diam.	Ill.	Elon.	Vis.	Rat.	Close To
21st	Sgr	18h 15m	-23° 40'	1.2	5"	92%	61° W	AM	*	
25th	Sgr	18h 27m	-23° 38'	1.2	5"	91%	61° W	AM	*	
29th	Sgr	18h 39m	-23° 32'	1.1	5"	91%	62° W	AM	*	

The Outer Planets

Planet	Date	Con.	R.A.	Dec.	Mag.	Diam.	Elon.	Vis.	Rat.	Close To
Jupiter	25th	Sgr	19h 21m	-22° 8'	-2.0	34"	48° W	AM	**	Saturn
Saturn	25th	Sgr	19h 59m	-20° 38'	0.7	15"	38° W	AM	**	
Uranus	25th	Ari	2h 5m	12° 15'	5.8	3"	53° E	PM	*	
Neptune	25th	Aqr	23h 17m	-5° 42'	8.0	2"	11° E	NV	N/A	Moon

Highlights

Date	Time (UT)	Event
23rd	15:33	New Moon. (Not visible.)
26th	01:39	Mercury is at inferior conjunction with the Sun. (Not visible.)
27th	10:27	The waxing crescent Moon is south of Venus. (Evening sky.)
	N/A	Good opportunity to see Earthshine on the waxing crescent Moon. (Evening sky.)
28th	10:30	The waxing crescent Moon is south of Uranus. (Evening sky.)

Planet Locations – February 25th

☉	☿	♀	♂	♃	♄
Sun	Mercury	Venus	Mars	Jupiter	Saturn

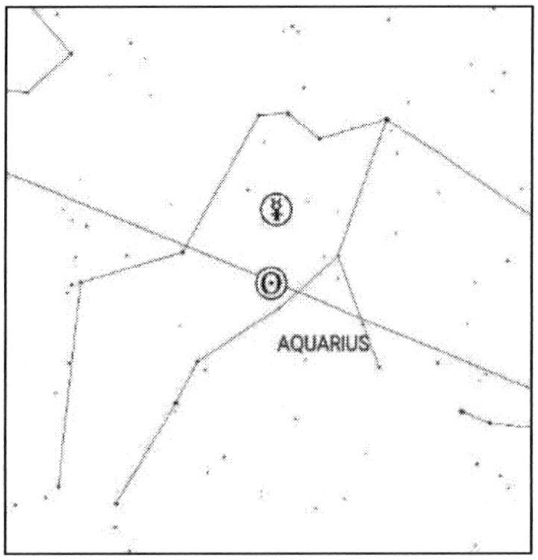

Sun

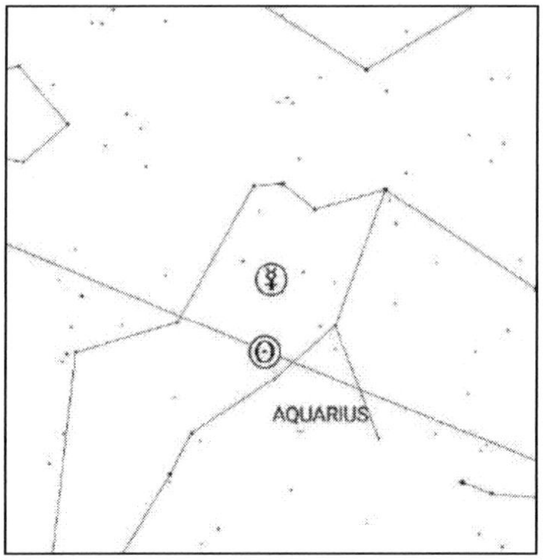

Mercury

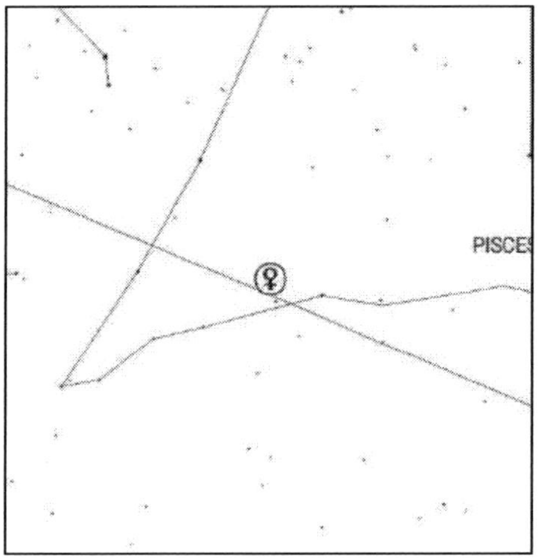

Venus

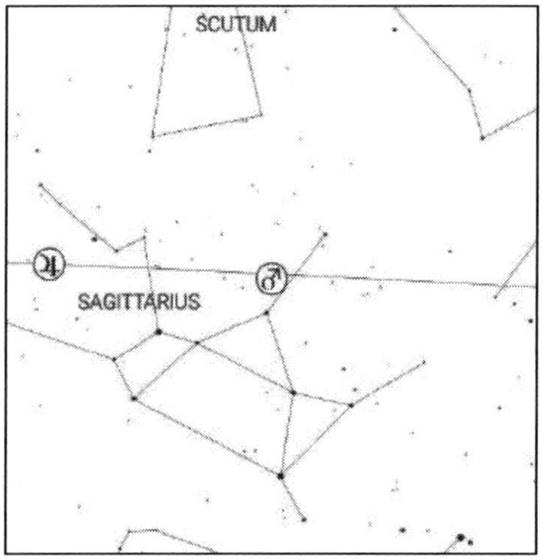

Mars

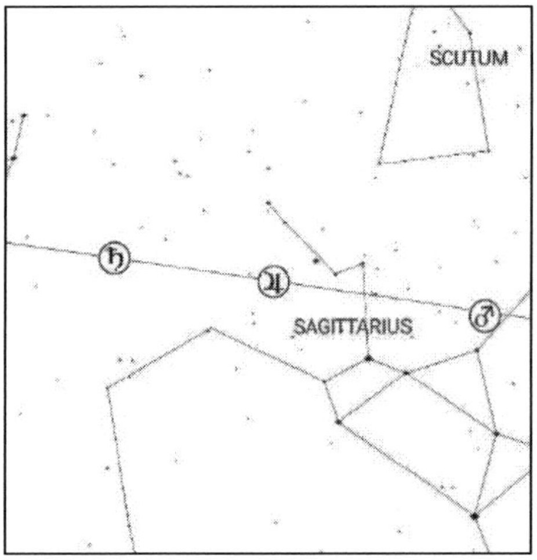

Jupiter

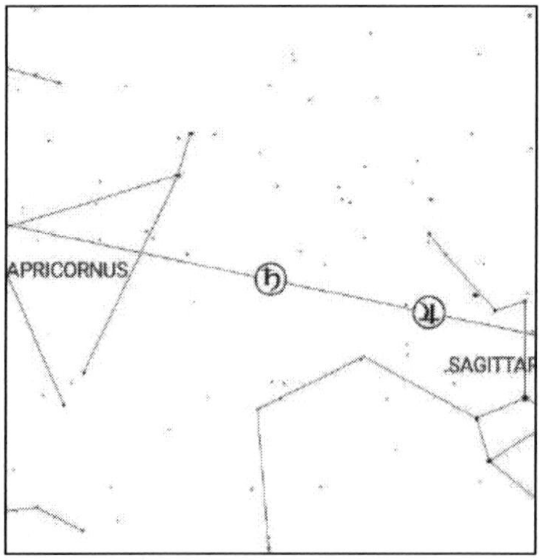

Saturn

February 25th – Evening Sky

Let's take a few moments to admire Orion (Ori), the Hunter, and his two faithful hounds. By drawing a curved line through the shoulders of Orion, we come to Procyon, the brightest star in Canis Minor (CMi), the Lesser Dog.

Next, follow the three stars of Orion's belt downward and you'll come to Sirius. The brightest star in Canis Major (CMa), the Greater Dog, it's also the brightest star in the sky.

This image depicts the sky looking south at about two hours after sunset.

February 26th – Evening Sky

Three days after turning new and the Moon returns to revisit Venus in the evening sky. You'll find them both hanging high over the western horizon tonight.

The Moon appears some way below Venus but it can move quickly from night to night. Come back tomorrow to see how the view has changed.

This image depicts the sky looking west at about an hour after sunset.

February 27th – Evening Sky

The month has nearly come to a close and the Moon has leapt closer to Venus. Last night it was some way below the planet, but this evening it appears to its left.

If you didn't see it yesterday, this is also another good opportunity to catch Earthshine illuminating the darkened surface of the Moon. It may still be there tomorrow, but will fade as the Moon moves toward first quarter on the 2nd.

This image depicts the sky looking west at about an hour after sunset.

March 1ˢᵗ to 10ᵗʰ, 2020

The Moon

| | 1ˢᵗ | 3ʳᵈ | 5ᵗʰ | 7ᵗʰ | 9ᵗʰ |

Date	Con	R.A.	Dec	Mag	Diam	Ill.	Elon.	Phase	Close To
1st	Tau	3h 38m	16° 10'	-9.5	30'	37%	72° E	+Cr	Pleiades
2nd	Tau	4h 28m	19° 26'	-10.0	30'	47%	83° E	FQ	Pleiades, Hyades, Aldebaran
3rd	Tau	5h 21m	21° 52'	-10.4	31'	57%	96° E	FQ	Aldebaran
4th	Gem	6h 18m	23° 14'	-10.8	31'	67%	109° E	+G	
5th	Gem	7h 16m	23° 19'	-11.2	32'	77%	123° E	+G	
6th	Cnc	8h 17m	21° 58'	-11.5	32'	85%	137° E	+G	Praesepe
7th	Cnc	9h 17m	19° 11'	-11.9	33'	93%	151° E	+G	Praesepe
8th	Leo	10h 17m	15° 4'	-12.3	33'	98%	165° E	FM	Regulus
9th	Leo	11h 15m	9° 55'	-12.6	33'	100%	179° E	FM	
10th	Vir	12h 12m	4° 6'	-12.4	33'	99%	168° W	FM	

Mercury and Venus

 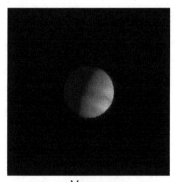

Mercury
5ᵗʰ

Venus
5ᵗʰ

Mercury

Date	Con.	R.A.	Dec.	Mag.	Diam.	Ill.	Elon.	Vis.	Rat.	Close To
1st	Aqr	22h 12m	-7° 31'	3.2	11"	5%	10° W	NV	N/A	
3rd	Aqr	22h 6m	-8° 26'	2.5	10"	9%	13° W	NV	N/A	
5th	Aqr	22h 2m	-9° 16'	1.9	10"	13%	16° W	AM	***	
7th	Aqr	21h 59m	-9° 58'	1.5	10"	18%	18° W	AM	***	
9th	Cap	21h 59m	-10° 33'	1.2	10"	23%	20° W	AM	***	

Venus

Date	Con.	R.A.	Dec.	Mag.	Diam.	Ill.	Elon.	Vis.	Rat.	Close To
1st	Psc	1h 34m	11° 7'	-4.2	19"	63%	41° E	PM	***	
3rd	Psc	1h 42m	12° 4'	-4.2	19"	62%	41° E	PM	***	
5th	Ari	1h 51m	12° 60'	-4.3	20"	61%	41° E	PM	***	Uranus
7th	Ari	1h 59m	13° 55'	-4.3	20"	60%	41° E	PM	***	Uranus
9th	Ari	2h 7m	14° 48'	-4.3	20"	59%	42° E	PM	***	Uranus

Mars and the Outer Planets

Mars
5th

Jupiter
5th

Saturn
5th

Mars

Date	Con.	R.A.	Dec.	Mag.	Diam.	Ill.	Elon.	Vis.	Rat.	Close To
1st	Sgr	18h 42m	-23° 30'	1.1	5"	91%	62° W	AM	*	
5th	Sgr	18h 54m	-23° 20'	1.1	6"	91%	63° W	AM	*	Jupiter
10th	Sgr	19h 9m	-23° 3'	1.0	6"	90%	64° W	AM	*	Jupiter

The Outer Planets

Planet	Date	Con.	R.A.	Dec.	Mag.	Diam.	Elon.	Vis.	Rat.	Close To
Jupiter	5th	Sgr	19h 28m	-21° 54'	-2.0	35"	55° W	AM	**	Mars, Saturn
Saturn	5th	Sgr	20h 2m	-20° 28'	0.7	16"	46° W	AM	**	
Uranus	5th	Ari	2h 7m	12° 22'	5.8	3"	45° E	PM	*	Venus
Neptune	5th	Aqr	23h 18m	-5° 34'	8.0	2"	3° E	NV	N/A	

Highlights

Date	Time (UT)	Event
1st	15:24	The nearly first quarter Moon is south of the Pleiades star cluster. (Taurus, evening sky.)
2nd	14:15	The almost first quarter Moon is north of the bright star Aldebaran. (Taurus, evening sky.)
	19:58	First Quarter Moon. (Evening sky.)
6th	20:44	The waxing gibbous Moon is north of the Praesepe star cluster. (Cancer, evening sky.)
8th	08:55	The nearly full Moon is north of the bright star Regulus. (Leo, evening sky.)
	23:15	Neptune is in conjunction with the Sun. (Not visible.)
9th	07:58	Mercury is stationary prior to resuming prograde motion. (Morning sky.)
	14:32	Venus is 2.4° north of Uranus. (Evening sky.)
	17:48	Full Moon. (Visible all night.)

Planet Locations – March 5th

☉	☿	♀	♂	♃	♄
Sun	Mercury	Venus	Mars	Jupiter	Saturn

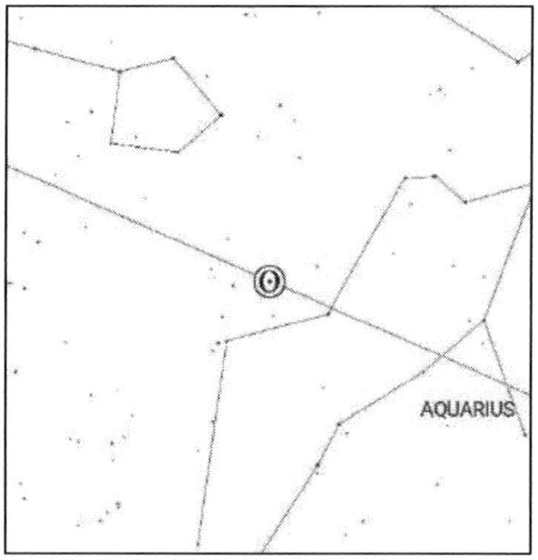

Sun

Mercury

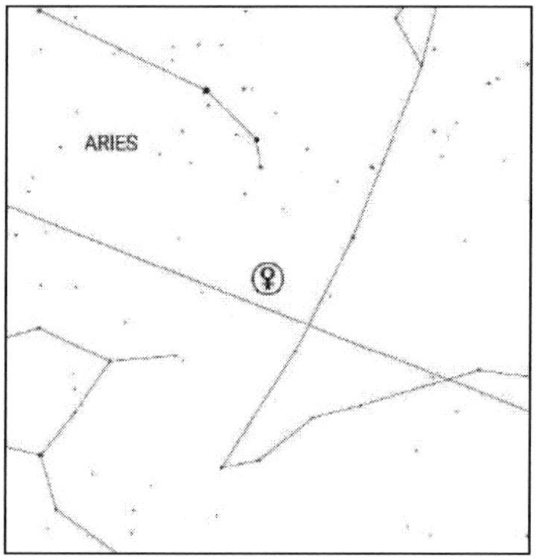

Venus

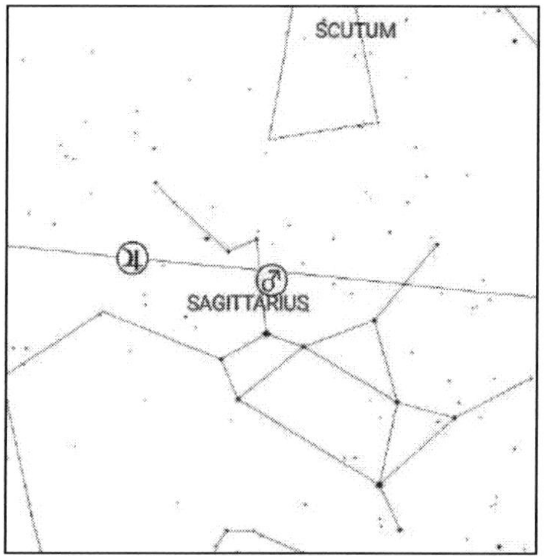

Mars

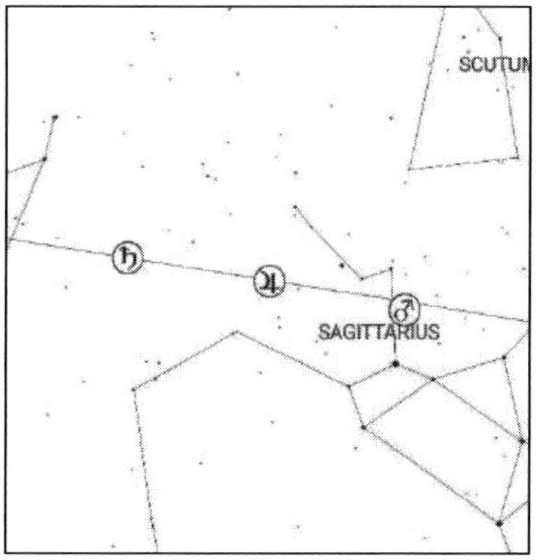

Jupiter

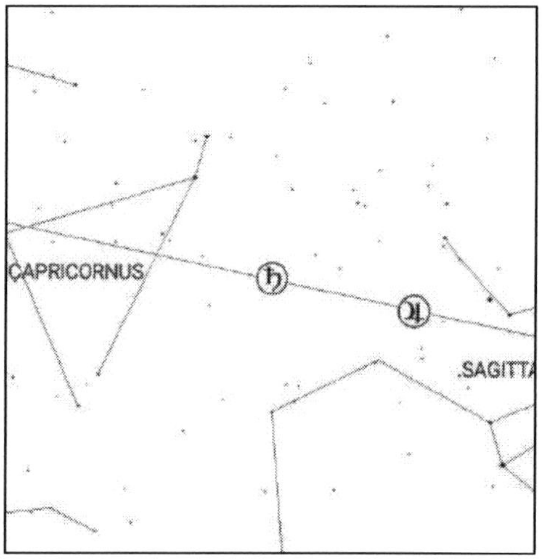

Saturn

March 1st – Evening Sky

The Moon reaches first quarter tomorrow and can be seen among the constellations of winter tonight.

It's currently approaching the stars of Orion (Ori) and Taurus (Tau), with orange Aldebaran to its upper left. The Pleiades lie to its right – can you see the stars of the cluster or does the Moon overwhelm their light?

This image depicts the sky looking south-west about two hours after sunset.

March 7th – Evening Sky

Still two days away from turning full, the Moon returns to Regulus, the brightest star in the spring constellation of Leo, the Lion.

At less than 80 light years away, the star is one our nearest neighbors. Its name means "little king" and is actually comprised of two pairs of stars orbiting one another.

This image depicts the sky looking east about an hour after sunset.

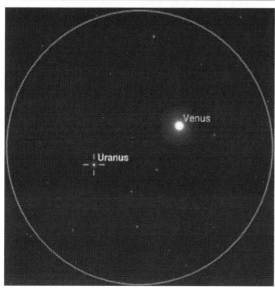

March 8th – Evening Sky

On January 27th, Venus passed close to distant Neptune in the night sky. There's another rare opportunity tonight as the planet now appears near to Neptune's planetary neighbor, Uranus.

Again, the pair will be easily seen within the same binocular view for a number of days, but this is the closest they will be this year.

This image depicts the view through 10x50 binoculars, looking west at about 9 p.m.

March 11th to 20th, 2020

The Moon

| 11th | 13th | 15th | 17th | 19th |

Date	Con	R.A.	Dec	Mag	Diam	Ill.	Elon.	Phase	Close To
11th	Vir	13h 8m	-1° 57'	-12.1	33'	95%	155° W	FM	Spica
12th	Vir	14h 3m	-7° 48'	-11.7	33'	89%	142° W	-G	Spica
13th	Lib	14h 59m	-13° 5'	-11.3	33'	80%	129° W	-G	
14th	Lib	15h 55m	-17° 29'	-10.9	32'	70%	116° W	-G	Antares
15th	Oph	16h 52m	-20° 46'	-10.5	32'	60%	103° W	LQ	Antares
16th	Sgr	17h 49m	-22° 48'	-10.1	31'	49%	89° W	LQ	
17th	Sgr	18h 45m	-23° 32'	-9.6	31'	39%	76° W	LQ	Mars
18th	Sgr	19h 41m	-23° 2'	-9.1	30'	29%	63° W	-Cr	Mars, Jupiter, Saturn
19th	Cap	20h 34m	-21° 24'	-8.4	30'	21%	51° W	-Cr	Saturn
20th	Cap	21h 24m	-18° 48'	-7.7	30'	13%	39° W	-Cr	

Mercury and Venus

Mercury
15th

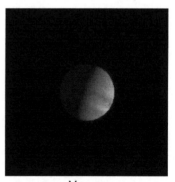

Venus
15th

Mercury

Date	Con.	R.A.	Dec.	Mag.	Diam.	Ill.	Elon.	Vis.	Rat.	Close To
11th	Aqr	22h 0m	-10° 58'	0.9	9"	28%	22° W	AM	***	
13th	Aqr	22h 2m	-11° 14'	0.7	9"	33%	23° W	AM	***	
15th	Aqr	22h 6m	-11° 22'	0.6	9"	37%	24° W	AM	****	
17th	Aqr	22h 11m	-11° 21'	0.5	8"	41%	25° W	AM	****	
19th	Aqr	22h 17m	-11° 13'	0.4	8"	45%	25° W	AM	****	

Venus

Date	Con.	R.A.	Dec.	Mag.	Diam.	Ill.	Elon.	Vis.	Rat.	Close To
11th	Ari	2h 15m	15° 41'	-4.3	21"	58%	42° E	PM	***	Uranus
13th	Ari	2h 23m	16° 32'	-4.3	21"	57%	42° E	PM	***	Uranus
15th	Ari	2h 31m	17° 21'	-4.3	21"	56%	42° E	PM	***	
17th	Ari	2h 39m	18° 9'	-4.3	22"	55%	42° E	PM	***	
19th	Ari	2h 47m	18° 55'	-4.3	22"	54%	43° E	PM	***	

Mars and the Outer Planets

Mars
15th

Jupiter
15th

Saturn
15th

Mars

Date	Con.	R.A.	Dec.	Mag.	Diam.	Ill.	Elon.	Vis.	Rat.	Close To
11th	Sgr	19h 12m	-22° 59'	1.0	6"	90%	64° W	AM	*	Jupiter
15th	Sgr	19h 24m	-22° 41'	1.0	6"	90%	65° W	AM	*	Jupiter
20th	Sgr	19h 39m	-22° 15'	0.9	6"	89%	66° W	AM	*	Jupiter, Saturn

The Outer Planets

Planet	Date	Con.	R.A.	Dec.	Mag.	Diam.	Elon.	Vis.	Rat.	Close To
Jupiter	15th	Sgr	19h 35m	-21° 40'	-2.0	35"	62° W	AM	**	Mars, Saturn
Saturn	15th	Sgr	20h 6m	-20° 18'	0.7	16"	54° W	AM	**	
Uranus	15th	Ari	2h 8m	12° 32'	5.9	3"	36° E	PM	*	Venus
Neptune	15th	Aqr	23h 19m	-5° 25'	8.0	2"	6° W	NV	N/A	

Highlights

Date	Time (UT)	Event
11th	18:11	The waning gibbous Moon is north of the bright star Spica. (Virgo, morning sky.)
15th	01:12	The nearly last quarter Moon is north of the bright star Antares. (Scorpius, morning sky.)
16th	09:35	Last Quarter Moon. (Morning sky.)
18th	08:38	The waning crescent Moon is south of Mars. (Morning sky.)
	11:20	The waning crescent Moon is south of Jupiter. (Morning sky.)
	23:06	The waning crescent Moon is south of Saturn. (Morning sky.)
20th	03:51	Spring equinox.
	06:14	Mars is 0.7° south of Jupiter. (Morning sky.)
	N/A	Good opportunity to see Earthshine on the waning crescent Moon. (Morning sky.)

Planet Locations – March 15ᵗʰ

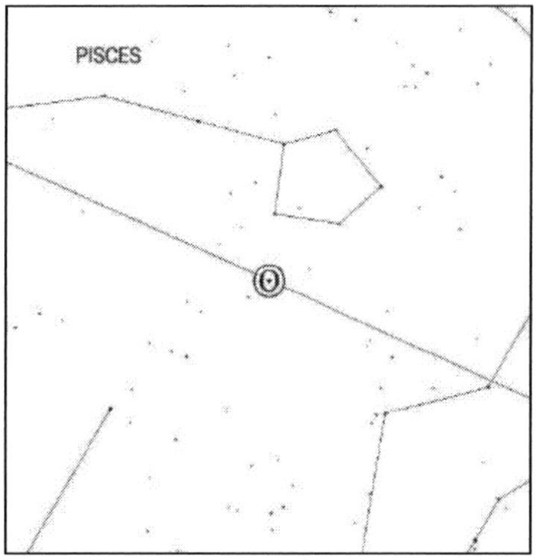

☉	☿	♀	♂	♃	♄
Sun	Mercury	Venus	Mars	Jupiter	Saturn

PISCES

☉

Sun

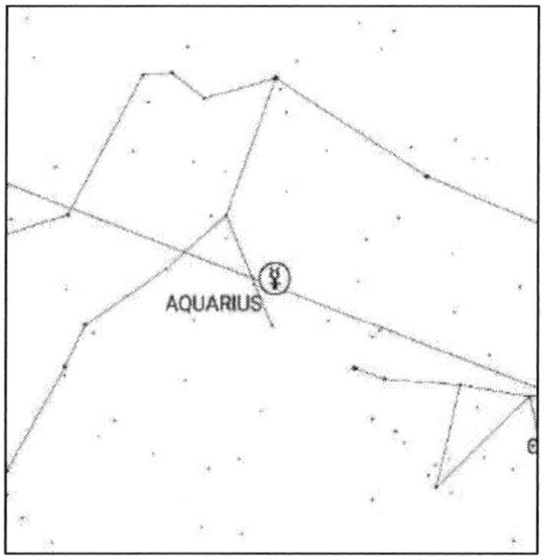

AQUARIUS ☿

Mercury

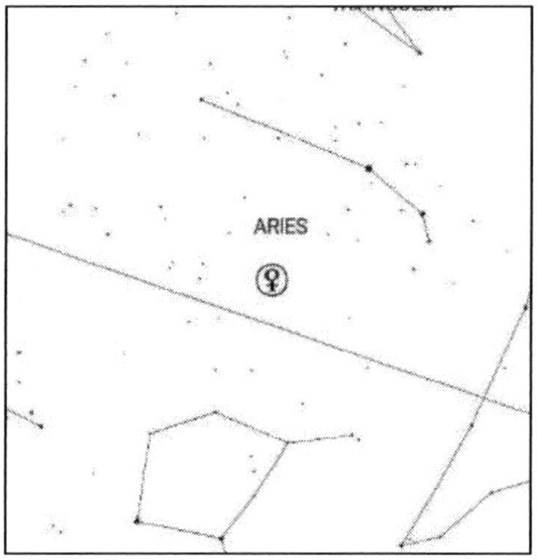

ARIES

♀

Venus

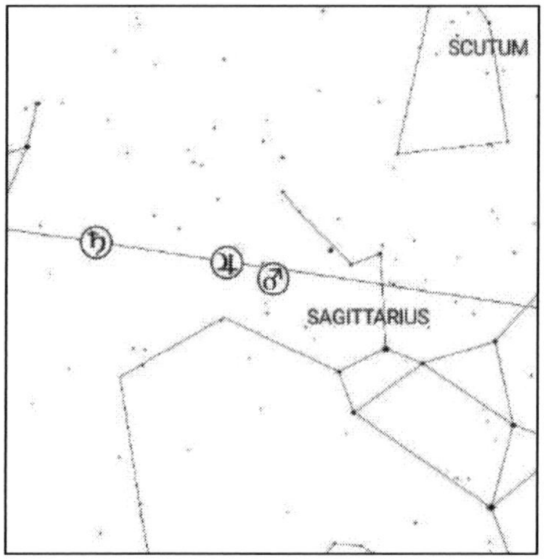

SCUTUM

♄ ♃ ♂

SAGITTARIUS

Mars

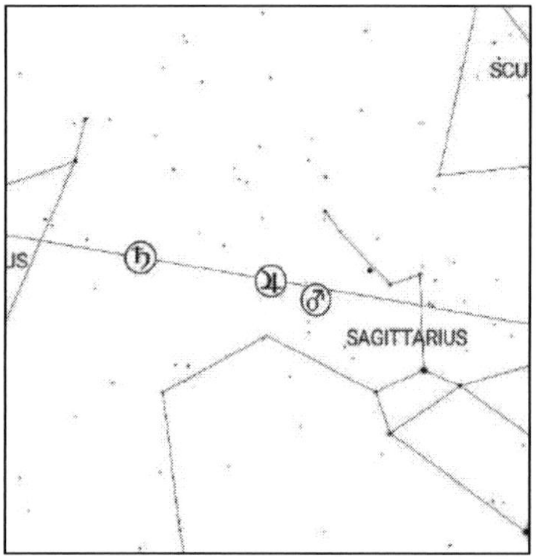

SCU

♄ ♃ ♂

SAGITTARIUS

Jupiter

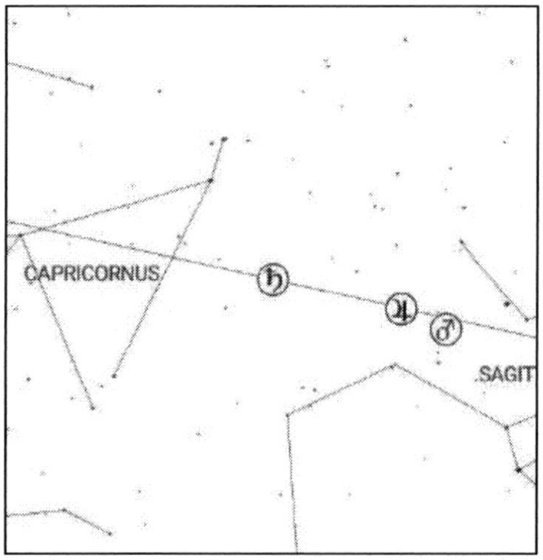

CAPRICORNUS ♄ ♃ ♂

.SAGIT

Saturn

March 15ᵗʰ – Morning Sky

Having left the stars of spring behind, the nearly last quarter Moon is now skimming the top of Scorpius (Sco), the Scorpion, a prominent summer constellation.

This morning it appears close to Antares before continuing its journey toward the planets Mars, Jupiter and Saturn in the pre-dawn sky.

This image depicts the sky looking south about an hour before sunrise.

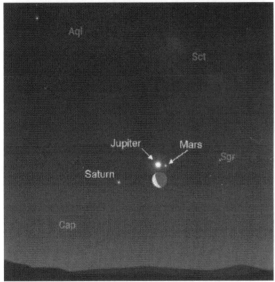

March 18ᵗʰ – Morning Sky

If you're an early riser, there's a nice grouping of planets waiting for you in the south-eastern sky.

Two months ago Mars appeared close to its rival, Antares, but has now caught up to Jupiter and Saturn in Sagittarius (Sgr), the Archer. It's to the right of Jupiter with the waning crescent Moon below the pair. Meanwhile, Saturn is a little way to the left. Come back tomorrow to see how far the Moon has moved in the intervening 24 hours.

This image depicts the sky looking east at about 45 minutes before sunrise.

March 20ᵗʰ – Morning Sky

The slender Moon now hangs low over the horizon this morning, but look at how Mars has moved closer to Jupiter.

This is the closest the pair will be and with less than the width of two full Moons between them, it should make for a very attractive sight. The red planet will catch up to Saturn in just ten days' time.

This image depicts the sky looking east at about 45 minutes before sunrise.

March 21st to 31st, 2020

The Moon

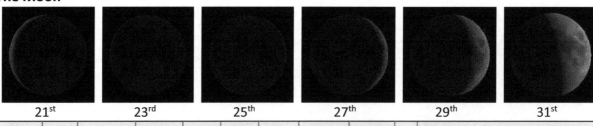

	21st		23rd		25th		27th		29th		31st

Date	Con	R.A.	Dec	Mag	Diam	Ill.	Elon.	Phase	Close To
21st	Aqr	22h 13m	-15° 25'	-7.0	30'	8%	28° W	NM	Mercury
22nd	Aqr	22h 59m	-11° 27'	-6.1	29'	3%	17° W	NM	Mercury, Neptune
23rd	Aqr	23h 44m	-7° 3'	-5.1	29'	1%	7° W	NM	Neptune
24th	Cet	0h 27m	-2° 24'	-4.4	29'	0%	3° E	NM	
25th	Cet	1h 11m	2° 19'	-5.3	29'	1%	13° E	NM	
26th	Psc	1h 54m	6° 59'	-6.2	29'	4%	23° E	NM	Uranus
27th	Ari	2h 39m	11° 25'	-7.1	30'	9%	33° E	NM	Venus, Uranus
28th	Tau	3h 26m	15° 27'	-7.9	30'	15%	44° E	+Cr	Venus, Pleiades
29th	Tau	4h 15m	18° 53'	-8.6	30'	22%	55° E	+Cr	Venus, Pleiades, Hyades, Aldebar
30th	Tau	5h 6m	21° 32'	-9.2	30'	31%	67° E	+Cr	Hyades, Aldebaran
31st	Tau	6h 0m	23° 13'	-9.7	31'	40%	80° E	FQ	

Mercury and Venus

Mercury
25th

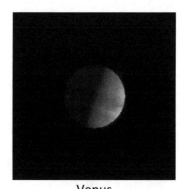

Venus
25th

Mercury

Date	Con.	R.A.	Dec.	Mag.	Diam.	Ill.	Elon.	Vis.	Rat.	Close To
21st	Aqr	22h 23m	-10° 57'	0.3	8"	48%	25° W	AM	****	Moon
23rd	Aqr	22h 31m	-10° 34'	0.3	7"	52%	25° W	AM	***	
25th	Aqr	22h 39m	-10° 4'	0.2	7"	55%	25° W	AM	***	
27th	Aqr	22h 47m	-9° 29'	0.2	7"	58%	25° W	AM	***	
29th	Aqr	22h 57m	-8° 47'	0.1	7"	60%	24° W	AM	***	
31st	Aqr	23h 6m	-7° 59'	0.1	7"	63%	24° W	AM	***	Neptune

Venus

Date	Con.	R.A.	Dec.	Mag.	Diam.	Ill.	Elon.	Vis.	Rat.	Close To
21st	Ari	2h 56m	19° 40'	-4.3	23"	53%	43° E	PM	***	
23rd	Ari	3h 4m	20° 23'	-4.4	23"	52%	43° E	PM	***	
25th	Ari	3h 11m	21° 5'	-4.4	24"	51%	43° E	PM	***	Pleiades
27th	Ari	3h 19m	21° 44'	-4.4	24"	50%	43° E	PM	***	Moon, Pleiades
29th	Ari	3h 27m	22° 22'	-4.4	25"	49%	43° E	PM	***	Moon, Pleiades
31st	Tau	3h 35m	22° 57'	-4.4	25"	47%	44° E	PM	***	Pleiades

Mars and the Outer Planets

Mars
25th

Jupiter
25th

Saturn
25th

Mars

Date	Con.	R.A.	Dec.	Mag.	Diam.	Ill.	Elon.	Vis.	Rat.	Close To
21st	Sgr	19h 42m	-22° 9'	0.9	6"	89%	66° W	AM	*	Jupiter, Saturn
25th	Sgr	19h 53m	-21° 43'	0.9	6"	89%	66° W	AM	*	Jupiter, Saturn
31st	Cap	20h 11m	-20° 59'	0.8	6"	88%	67° W	AM	**	Jupiter, Saturn

The Outer Planets

Planet	Date	Con.	R.A.	Dec.	Mag.	Diam.	Elon.	Vis.	Rat.	Close To
Jupiter	25th	Sgr	19h 41m	-21° 26'	-2.1	36"	69° W	AM	**	Mars, Saturn
Saturn	25th	Cap	20h 9m	-20° 9'	0.7	16"	62° W	AM	**	Mars
Uranus	25th	Ari	2h 10m	12° 42'	5.9	3"	28° E	PM	*	
Neptune	25th	Aqr	23h 21m	-5° 17'	8.0	2"	15° W	NV	N/A	

Highlights

Date	Time (UT)	Event
21st	19:00	The waning crescent Moon is south of Mercury. (Morning sky.)
24th	02:03	Mercury is at greatest western elongation from the Sun. (Morning sky.)
	09:29	New Moon. (Not visible.)
	22:00	Venus is at greatest eastern elongation from the Sun. (Evening sky.)
26th	21:55	The waxing crescent Moon is south of Uranus. (Evening sky.)
27th	N/A	Good opportunity to see Earthshine on the waxing crescent Moon. (Evening sky.)
28th	09:11	The waxing crescent Moon is south of Venus. (Evening sky.)
	23:28	The waxing crescent Moon is south of the Pleaides star cluster. (Taurus, evening sky.)
29th	23:09	The waxing crescent Moon is north of the bright star Aldebaran. (Taurus, evening sky.)
31st	10:50	Mars is 0.9° south of Saturn. (Morning sky.)

Planet Locations – March 25th

Sun	Moon	Mercury	Venus	Mars	Jupiter	Saturn

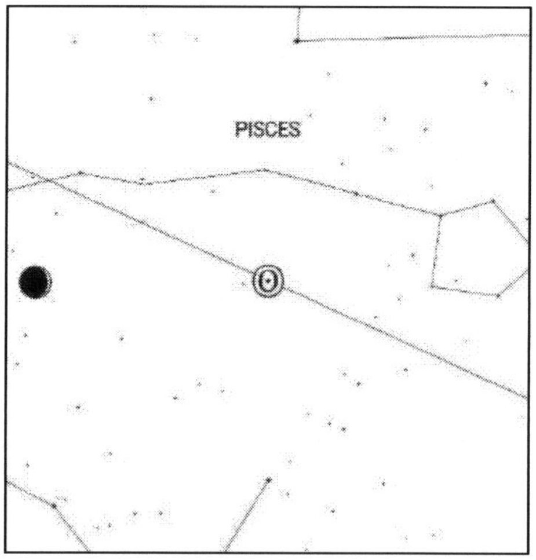

PISCES

Sun

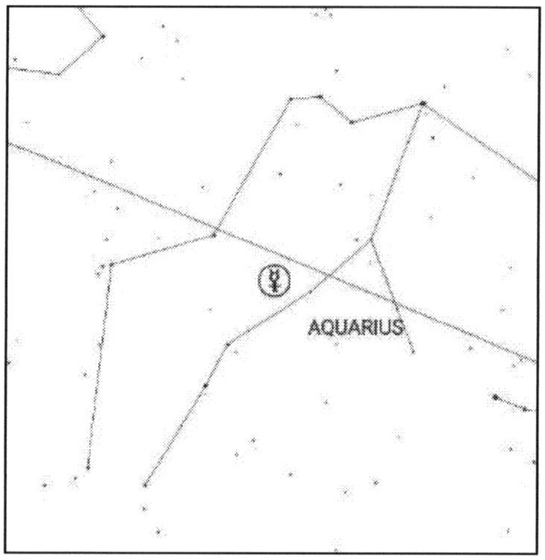

AQUARIUS

Mercury

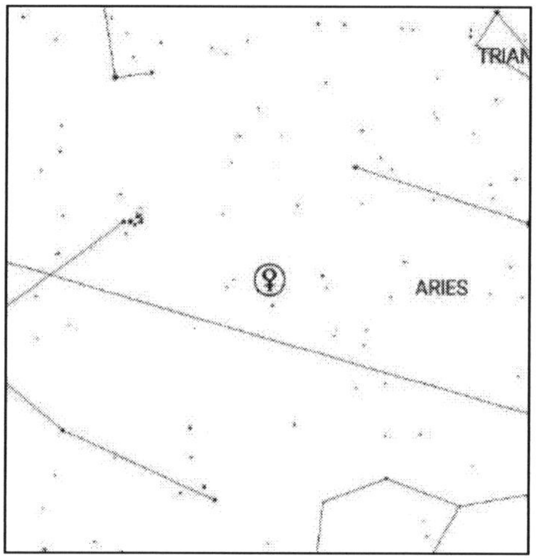

TRIAN

ARIES

Venus

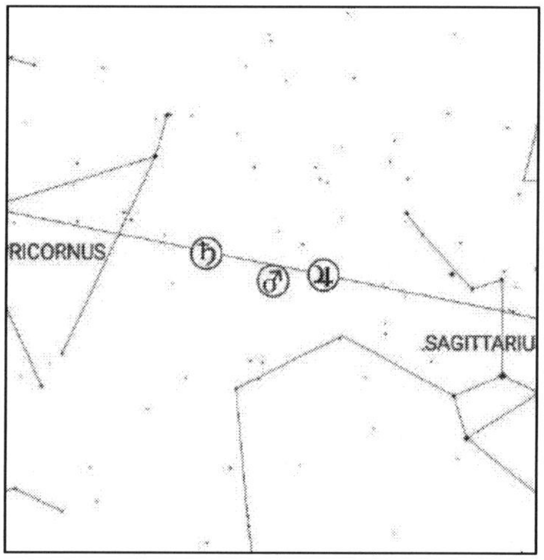

RICORNUS

SAGITTARIU

Mars

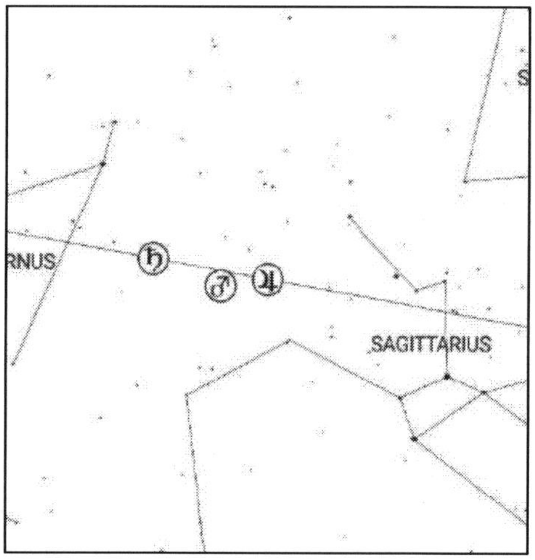

RNUS

SAGITTARIUS

Jupiter

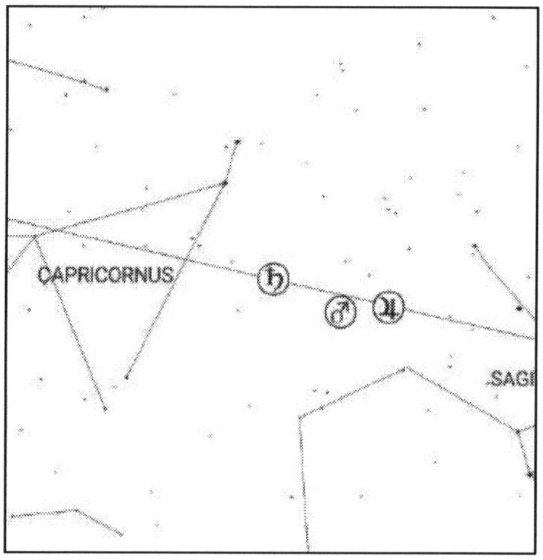

CAPRICORNUS

SAGI

Saturn

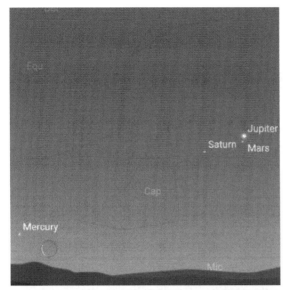

March 21st – Morning Sky

Mercury is edging away from the Sun in the pre-dawn sky, but might be a little tricky to identify against the brightening twilight.

Fortunately, the slender crescent Moon appears to its lower right this morning, making it a little easier to spot. Meanwhile, Mars has now slipped past Jupiter and now continues on toward Saturn.

This image depicts the sky looking east at about 15 minutes before sunrise.

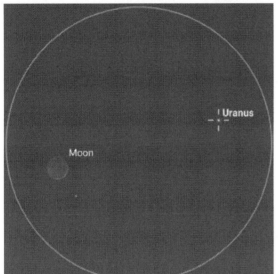

March 26th – Evening Sky

Two months ago the nearly first quarter Moon passed close to Uranus in the evening sky. If you missed it then, you have another chance tonight, but the sky will be brighter and the planet harder to spot.

Binoculars will show the pair within the same field of view. Uranus should look like a pale blue star, to the right of the thin lunar crescent.

This image depicts the view through 10x50 binoculars, looking west at about an hour after sunset.

March 31st – Morning Sky

Eleven days ago, Mars was closest to Jupiter in the pre-dawn sky. Now it's passing by Saturn.

The gap is slightly wider than the Mars-Jupiter pairing, but the color contrast should still be very apparent. Come back tomorrow to see how much further Mars has moved along.

This image depicts the view looking east at about 45 minutes before sunrise.

April 1st to 10th, 2020

The Moon

| | 1st | | 3rd | | 5th | | 7th | | 9th |

Date	Con	R.A.	Dec	Mag	Diam	Ill.	Elon.	Phase	Close To
1st	Gem	6h 57m	23° 42'	-10.2	31'	51%	93° E	FQ	
2nd	Gem	7h 55m	22° 52'	-10.6	32'	61%	107° E	FQ	Praesepe
3rd	Cnc	8h 53m	20° 40'	-11.0	32'	72%	120° E	+G	Praesepe
4th	Leo	9h 52m	17° 8'	-11.4	33'	82%	134° E	+G	Regulus
5th	Leo	10h 49m	12° 27'	-11.8	33'	90%	148° E	+G	Regulus
6th	Vir	11h 46m	6° 54'	-12.1	33'	96%	161° E	FM	
7th	Vir	12h 42m	0° 51'	-12.5	33'	99%	174° E	FM	Spica
8th	Vir	13h 38m	-5° 16'	-12.5	33'	100%	173° W	FM	Spica
9th	Lib	14h 35m	-11° 1'	-12.2	33'	97%	160° W	FM	
10th	Lib	15h 32m	-16° 1'	-11.8	33'	92%	146° W	-G	

Mercury and Venus

 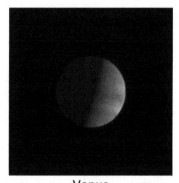

Mercury
5th

Venus
5th

Mercury

Date	Con.	R.A.	Dec.	Mag.	Diam.	Ill.	Elon.	Vis.	Rat.	Close To
1st	Aqr	23h 11m	-7° 34'	0.1	7"	64%	23° W	AM	***	Neptune
3rd	Aqr	23h 21m	-6° 38'	0.0	6"	67%	23° W	AM	***	Neptune
5th	Aqr	23h 32m	-5° 37'	0.0	6"	69%	22° W	AM	***	Neptune
7th	Aqr	23h 43m	-4° 32'	-0.1	6"	71%	21° W	AM	***	
9th	Aqr	23h 54m	-3° 22'	-0.1	6"	74%	20° W	AM	***	

Venus

Date	Con.	R.A.	Dec.	Mag.	Diam.	Ill.	Elon.	Vis.	Rat.	Close To
1st	Tau	3h 39m	23° 14'	-4.4	26"	47%	44° E	PM	***	Pleiades
3rd	Tau	3h 46m	23° 47'	-4.4	26"	46%	44° E	PM	***	Pleiades
5th	Tau	3h 54m	24° 18'	-4.4	27"	44%	44° E	PM	***	Pleiades, Hyades
7th	Tau	4h 1m	24° 46'	-4.4	28"	43%	44° E	PM	***	Pleiades, Hyades, Aldebaran
9th	Tau	4h 9m	25° 13'	-4.5	28"	42%	44° E	PM	****	Pleiades, Hyades, Aldebaran

Mars and the Outer Planets

Mars
5th

Jupiter
5th

Saturn
5th

Mars

Date	Con.	R.A.	Dec.	Mag.	Diam.	Ill.	Elon.	Vis.	Rat.	Close To
1st	Cap	20h 14m	-20° 51'	0.8	6"	88%	68° W	AM	**	Jupiter, Saturn
5th	Cap	20h 26m	-20° 18'	0.7	7"	88%	68° W	AM	**	Jupiter, Saturn
10th	Cap	20h 40m	-19° 32'	0.7	7"	88%	69° W	AM	**	Saturn

The Outer Planets

Planet	Date	Con.	R.A.	Dec.	Mag.	Diam.	Elon.	Vis.	Rat.	Close To
Jupiter	5th	Sgr	19h 47m	-21° 13'	-2.2	38"	78° W	AM	**	Mars, Saturn
Saturn	5th	Cap	20h 12m	-20° 1'	0.7	16"	72° W	AM	**	Mars
Uranus	5th	Ari	2h 13m	12° 54'	5.9	3"	18° E	PM	*	
Neptune	5th	Aqr	23h 22m	-5° 8'	8.0	2"	24° W	AM	*	Mercury

Highlights

Date	Time (UT)	Event
1st	10:22	First Quarter Moon. (Evening sky.)
3rd	06:43	The waxing gibbous Moon is north of the Praesepe star cluster. (Cancer, evening sky.)
	11:59	Venus is 0.3° south of the Pleiades star cluster. (Taurus, evening sky.)
	15:18	Mercury is 1.4° south of Neptune. (Morning sky.)
4th	17:41	The waxing gibbous Moon is north of the bright star Regulus. (Leo, evening sky.)
8th	02:36	Full Moon. (Visible all night.)
	07:29	The full Moon is north of the bright star Spica. (Virgo, visible all night.)

Planet Locations – April 5th

Sun	Mercury	Venus	Mars	Jupiter	Saturn

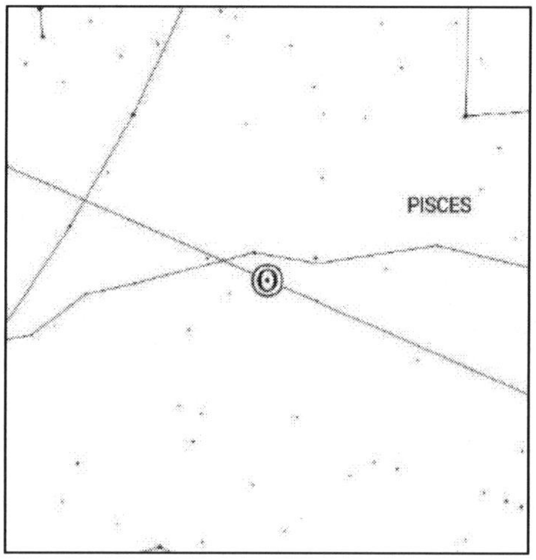

Sun

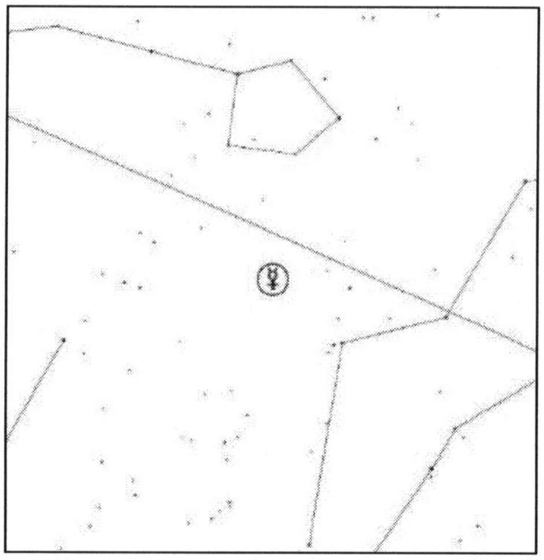

Mercury

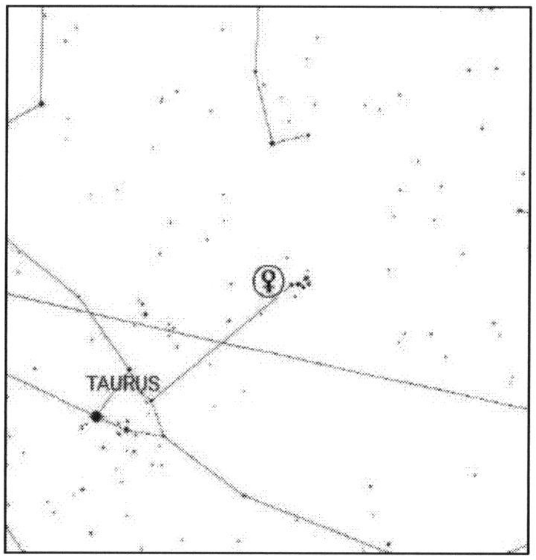

Venus

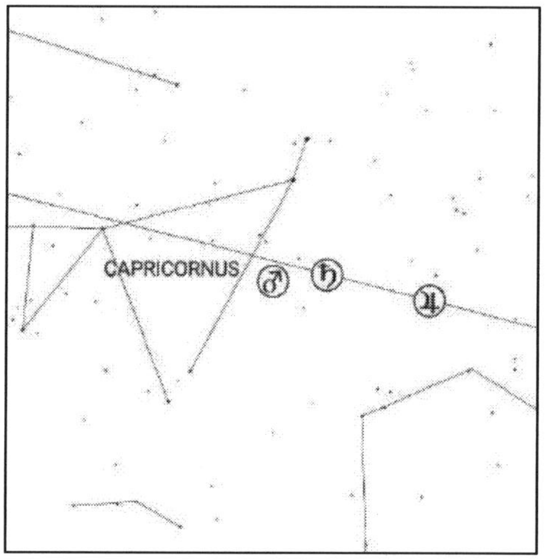

Mars

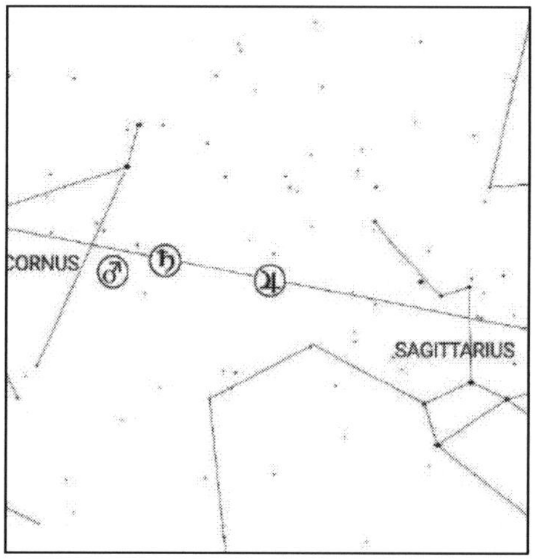

Jupiter

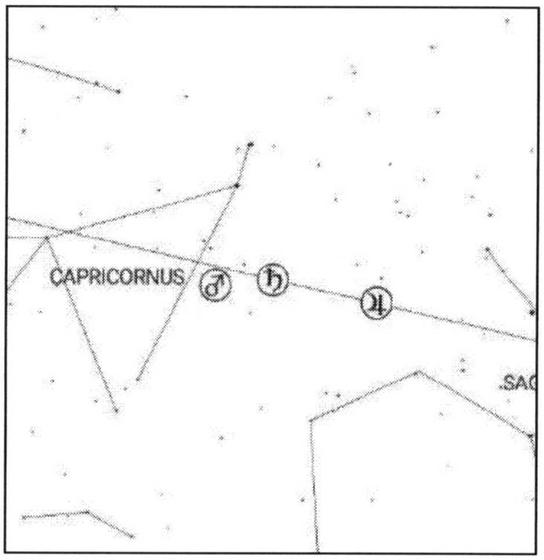

Saturn

April 3rd – Evening Sky

This evening presents a wonderful opportunity to see the planet Venus among the stars of the Pleiades. The planet and the cluster will easily fit within the same binocular field of view and could also be quite stunning through a telescope with a very low power eyepiece.

Not to be missed – but if clouds deny you tonight, they'll still be close tomorrow.

This image depicts the view through 10x50 binoculars, looking west about two hours after sunset.

April 4th – Evening Sky

The Moon is currently in its waxing gibbous phase and is roughly 85% illuminated. As such, it should look nearly, but not quite full.

Tonight it returns to Regulus, the brightest star in the constellation of Leo, the Lion. Over the next few nights it will leave the star and constellation behind and venture into neighboring Virgo (Vir).

This image depicts the sky looking south-east at about two hours after sunset.

April 7th – Evening Sky

If you go by Universal Time, the Moon turns full in the early hours of the 8th. However, for North American observers, this means the full Moon actually occurs in the evening of the 7th.

So this is a good opportunity to see the Moon 100% illuminated. It's currently moving through Virgo (Vir), which isn't a particularly bright constellation, but its brightest star Spica should still be easily seen.

This image depicts the sky looking south-east at about two hours after sunset.

April 11th to 20th, 2020

The Moon

| 11th | 13th | 15th | 17th | 19th |

Date	Con	R.A.	Dec	Mag	Diam	Ill.	Elon.	Phase	Close To
11th	Oph	16h 31m	-19° 55'	-11.5	32'	84%	133° W	-G	Antares
12th	Oph	17h 30m	-22° 31'	-11.1	32'	75%	119° W	-G	
13th	Sgr	18h 28m	-23° 43'	-10.7	31'	65%	105° W	LQ	
14th	Sgr	19h 26m	-23° 33'	-10.3	31'	55%	92° W	LQ	Jupiter
15th	Cap	20h 20m	-22° 10'	-9.9	30'	45%	79° W	LQ	Mars, Jupiter, Saturn
16th	Cap	21h 12m	-19° 45'	-9.4	30'	35%	67° W	-Cr	Mars
17th	Aqr	22h 1m	-16° 30'	-8.8	30'	26%	55° W	-Cr	
18th	Aqr	22h 48m	-12° 37'	-8.2	30'	18%	45° W	-Cr	Neptune
19th	Aqr	23h 33m	-8° 17'	-7.5	29'	11%	34° W	NM	Neptune
20th	Psc	0h 17m	-3° 39'	-6.7	29'	6%	24° W	NM	Mercury

Mercury and Venus

Mercury
15th

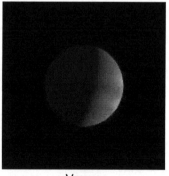

Venus
15th

Mercury

Date	Con.	R.A.	Dec.	Mag.	Diam.	Ill.	Elon.	Vis.	Rat.	Close To
11th	Psc	0h 5m	-2° 7'	-0.2	6"	76%	19° W	AM	**	
13th	Psc	0h 17m	0° 48'	-0.3	6"	78%	18° W	AM	**	
15th	Cet	0h 29m	0° 35'	-0.4	6"	81%	17° W	AM	**	
17th	Cet	0h 42m	2° 2'	-0.5	5"	83%	15° W	AM	**	
19th	Psc	0h 55m	3° 32'	-0.6	5"	86%	14° W	NV	N/A	

Venus

Date	Con.	R.A.	Dec.	Mag.	Diam.	Ill.	Elon.	Vis.	Rat.	Close To
11th	Tau	4h 16m	25° 37'	-4.5	29"	40%	44° E	PM	****	Pleiades, Hyades, Aldebaran
13th	Tau	4h 22m	25° 60'	-4.5	30"	39%	44° E	PM	****	Pleiades, Hyades, Aldebaran
15th	Tau	4h 29m	26° 20'	-4.5	31"	38%	43° E	PM	****	Hyades, Aldebaran
17th	Tau	4h 36m	26° 38'	-4.5	32"	36%	43° E	PM	****	Hyades, Aldebaran
19th	Tau	4h 42m	26° 54'	-4.5	33"	35%	43° E	PM	****	Hyades, Aldebaran

Mars and the Outer Planets

Mars
15th

Jupiter
15th

Saturn
15th

Mars

Date	Con.	R.A.	Dec.	Mag.	Diam.	Ill.	Elon.	Vis.	Rat.	Close To
11th	Cap	20h 43m	-19° 23'	0.7	7"	88%	70° W	AM	**	Saturn
15th	Cap	20h 54m	-18° 43'	0.6	7"	87%	70° W	AM	**	Moon
20th	Cap	21h 9m	-17° 50'	0.5	7"	87%	71° W	AM	**	

The Outer Planets

Planet	Date	Con.	R.A.	Dec.	Mag.	Diam.	Elon.	Vis.	Rat.	Close To
Jupiter	15th	Sgr	19h 51m	-21° 3'	-2.2	39"	86° W	AM	***	Moon, Saturn
Saturn	15th	Cap	20h 14m	-19° 55'	0.6	17"	80° W	AM	***	Moon
Uranus	15th	Ari	2h 15m	13° 5'	5.9	3"	10° E	NV	N/A	
Neptune	15th	Aqr	23h 23m	-4° 60'	7.9	2"	33° W	AM	*	

Highlights

Date	Time (UT)	Event
11th	12:07	The waning gibbous Moon is north of the bright star Antares. (Scorpius, morning sky.)
14th	20:25	The almost last quarter Moon is south of the dwarf planet Pluto. (Morning sky.)
	22:02	The almost last quarter Moon is south of Jupiter. (Morning sky.)
	22:57	Last Quarter Moon. (Morning sky.)
15th	10:23	The just-past last quarter Moon is south of Saturn. (Morning sky.)
16th	03:25	The waning crescent Moon is south of Mars. (Morning sky.)
19th	06:04	The waning crescent Moon is south of Neptune. (Morning sky.)
	N/A	Good opportunity to see Earthshine on the waning crescent Moon. (Morning sky.)

Planet Locations – April 15th

☉	☽	☿	♀	♂	♃	♄
Sun	Moon	Mercury	Venus	Mars	Jupiter	Saturn

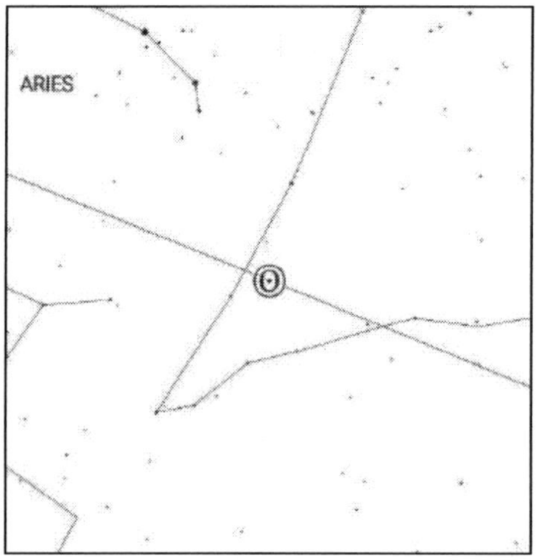

Sun

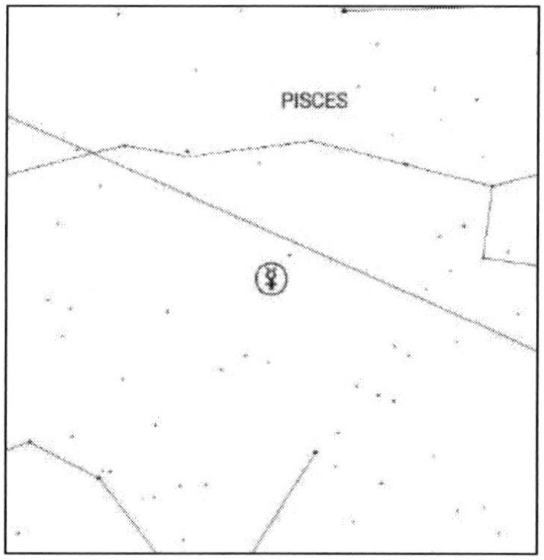

Mercury

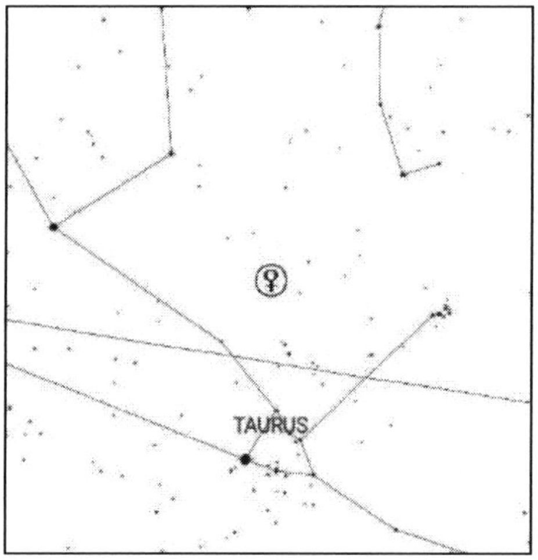

Venus

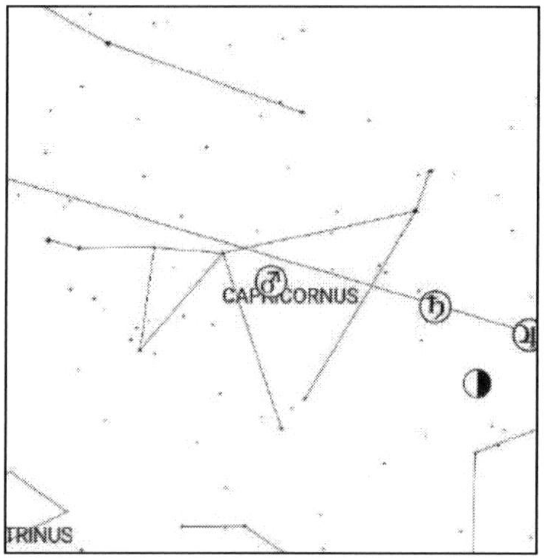

Mars

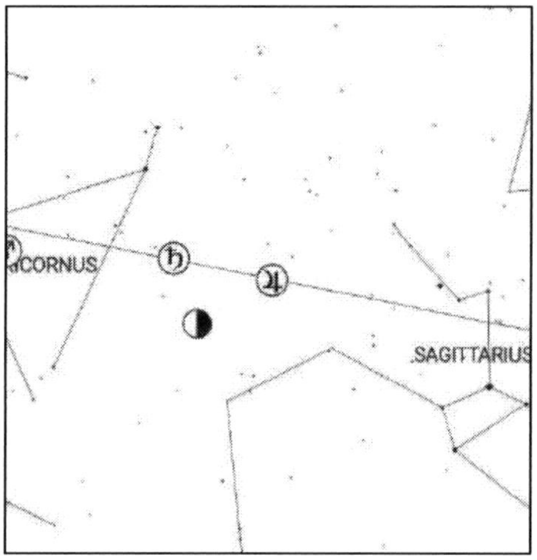

Jupiter

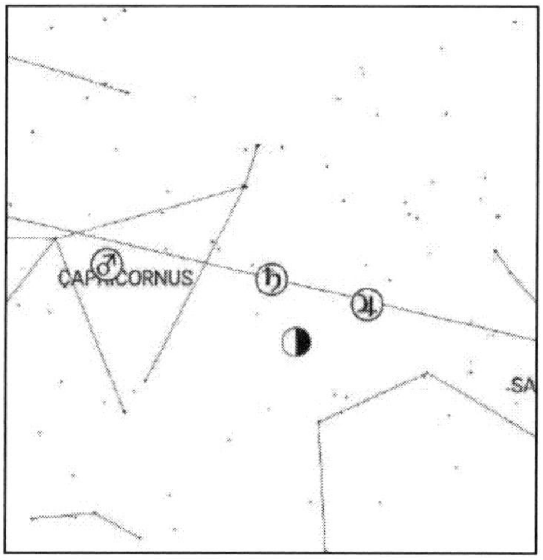

Saturn

April 14th – Morning Sky

It's been a week since the Moon turned full, which means it's now at the last quarter phase and has moved into the pre-dawn sky.

Over the next few mornings, it'll pass by the trio of outer planets that can be found here. First up is Jupiter, our solar system's largest world. It's not a particularly close conjunction, but the view tomorrow should be better.

This image depicts the sky looking south-east at about an hour before sunrise.

April 15th – Morning Sky

If you missed the conjunction between the Moon and Jupiter yesterday, you can still enjoy the view this morning.

The Moon now appears directly below Saturn, with Jupiter to the right and Mars to the left.

This image depicts the sky looking south-east at about an hour before sunrise.

April 16th – Morning Sky

Here's the last of the early morning lunar encounters for April. The Moon is now visibly slimming down to a crescent and appears to the lower left of Mars.

Compare the brightness of Mars to Saturn – which one is fainter? Or are they roughly the same?

This image depicts the sky looking south-east at about an hour before sunrise.

April 21ˢᵗ to 30ᵗʰ, 2020

The Moon

| | 21ˢᵗ | | 23ʳᵈ | | 25ᵗʰ | | 27ᵗʰ | | 29ᵗʰ | |

Date	Con	R.A.	Dec	Mag	Diam	Ill.	Elon.	Phase	Close To
21st	Cet	1h 0m	1° 7'	-5.8	29'	2%	15° W	NM	Mercury
22nd	Psc	1h 43m	5° 51'	-4.8	29'	0%	5° W	NM	Mercury, Uranus
23rd	Cet	2h 28m	10° 25'	-4.5	30'	0%	6° E	NM	Uranus
24th	Ari	3h 14m	14° 38'	-5.6	30'	2%	16° E	NM	Pleiades
25th	Tau	4h 3m	18° 17'	-6.5	30'	5%	27° E	NM	Pleiades, Hyades, Aldebaran
26th	Tau	4h 54m	21° 11'	-7.4	30'	11%	39° E	NM	Venus, Hyades, Aldebaran
27th	Tau	5h 47m	23° 8'	-8.2	30'	17%	52° E	+Cr	Venus
28th	Gem	6h 43m	23° 56'	-8.8	31'	26%	64° E	+Cr	
29th	Gem	7h 39m	23° 27'	-9.4	31'	36%	78° E	+Cr	
30th	Cnc	8h 37m	21° 40'	-10.0	32'	46%	91° E	FQ	Praesepe

Mercury and Venus

Mercury
25ᵗʰ

Venus
25ᵗʰ

Mercury

Date	Con.	R.A.	Dec.	Mag.	Diam.	Ill.	Elon.	Vis.	Rat.	Close To
21st	Psc	1h 8m	5° 6'	-0.8	5"	88%	13° W	NV	N/A	Moon
23rd	Psc	1h 22m	6° 42'	-0.9	5"	91%	11° W	NV	N/A	
25th	Psc	1h 36m	8° 21'	-1.1	5"	93%	9° W	NV	N/A	
27th	Psc	1h 50m	10° 1'	-1.3	5"	95%	8° W	NV	N/A	
29th	Ari	2h 6m	11° 42'	-1.5	5"	97%	6° W	NV	N/A	Uranus

Venus

Date	Con.	R.A.	Dec.	Mag.	Diam.	Ill.	Elon.	Vis.	Rat.	Close To
21st	Tau	4h 48m	27° 8'	-4.5	34"	33%	42° E	PM	****	Hyades, Aldebaran
23rd	Tau	4h 53m	27° 20'	-4.5	35"	31%	42° E	PM	****	Hyades, Aldebaran
25th	Tau	4h 58m	27° 30'	-4.5	36"	30%	41° E	PM	****	Hyades, Aldebaran
27th	Tau	5h 3m	27° 38'	-4.5	37"	28%	41° E	PM	****	Moon, Hyades, Aldebaran
29th	Tau	5h 7m	27° 44'	-4.5	38"	26%	40° E	PM	****	Aldebaran

Mars and the Outer Planets

Mars
25th

Jupiter
25th

Saturn
25th

Mars

Date	Con.	R.A.	Dec.	Mag.	Diam.	Ill.	Elon.	Vis.	Rat.	Close To
21st	Cap	21h 11m	-17° 40'	0.5	7"	87%	72° W	AM	**	
25th	Cap	21h 23m	-16° 54'	0.5	7"	87%	73° W	AM	**	
30th	Cap	21h 36m	-15° 56'	0.4	8"	86%	74° W	AM	**	

The Outer Planets

Planet	Date	Con.	R.A.	Dec.	Mag.	Diam.	Elon.	Vis.	Rat.	Close To
Jupiter	25th	Sgr	19h 54m	-20° 56'	-2.3	40"	95° W	AM	***	Saturn
Saturn	25th	Cap	20h 16m	-19° 52'	0.6	17"	89° W	AM	***	
Uranus	25th	Ari	2h 17m	13° 17'	5.9	3"	1° E	NV	N/A	
Neptune	25th	Aqr	23h 25m	-4° 53'	7.9	2"	42° W	AM	**	

Highlights

Date	Time (UT)	Event
22nd	N/A	The Lyrid meteor shower is at its maximum. (ZHR: 18)
23rd	02:26	New Moon. (Not visible.)
25th	03:28	The waxing crescent Moon is south of the Pleiades star cluster. (Evening sky.)
	21:49	Dwarf planet Pluto is stationary prior to beginning retrograde motion. (Morning sky.)
26th	03:14	The waxing crescent Moon is north of the bright star Aldebaran. (Evening sky.)
	13:10	Uranus is in conjunction with the Sun. (Not visible.)
	15:49	The waxing crescent Moon is south of Venus. (Evening sky.)
	N/A	Good opportunity to see Earthshine on the waxing crescent Moon. (Evening sky.)
30th	12:03	The almost first quarter Moon is north of the Praesepe star cluster. (Evening sky.)
	20:39	First Quarter Moon.

Planet Locations – April 25th

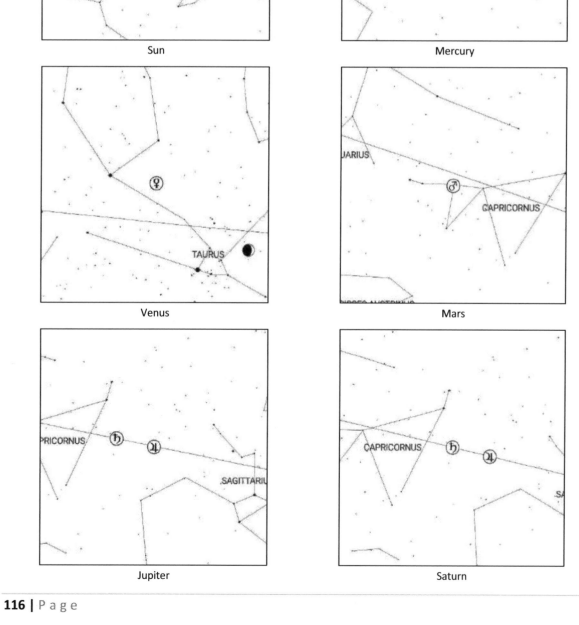

April 22nd to 23rd – Evening & Morning Sky

The Lyrid meteor shower reaches its peak in the early hours of this morning. A moderate shower, under ideal conditions you could expect to see about 18 shooting stars every hour.

They appear to originate from close to the constellation of Lyra (Lyr), so you'll need to look out for the bright star Vega in the north-east. Fortunately, the Moon turns new today and won't be an interference.

This image depicts the sky looking north-east at about 1:00 a.m.

April 25th – Evening Sky

The Moon turned new a few days ago and should now become visible again in the evening sky. Tonight it appears among the stars of Taurus (Tau), the Bull.

Look out for brilliant Venus a little way above it. Make a mental note of their positions and see how the scene has changed tomorrow.

This image depicts the sky looking west at about an hour a half after sunset.

April 26th – Evening Sky

Now the Moon and Venus are pretty much level and hover at the same height above the western horizon.

The stars of winter, now setting in the spring sky, surround the pair and Earthshine should be easily visible upon the Moon's surface.

This image depicts the sky looking west at about an hour a half after sunset.

May 1st to 10th, 2020

The Moon

1st	3rd	5th	7th	9th

Date	Con	R.A.	Dec	Mag	Diam	Ill.	Elon.	Phase	Close To
1st	Leo	9h 33m	18° 36'	-10.4	32'	57%	104° E	FQ	Regulus
2nd	Leo	10h 29m	14° 24'	-10.9	32'	68%	117° E	+G	Regulus
3rd	Leo	11h 24m	9° 16'	-11.3	33'	78%	130° E	+G	
4th	Vir	12h 19m	3° 31'	-11.6	33'	87%	143° E	+G	
5th	Vir	13h 13m	-2° 33'	-12.0	33'	94%	155° E	+G	Spica
6th	Vir	14h 9m	-8° 30'	-12.4	33'	99%	168° E	FM	Spica
7th	Lib	15h 6m	-13° 56'	-12.6	33'	100%	178° W	FM	
8th	Sco	16h 5m	-18° 27'	-12.3	33'	98%	165° W	FM	Antares
9th	Oph	17h 5m	-21° 44'	-12.0	32'	94%	151° W	-G	Antares
10th	Sgr	18h 5m	-23° 35'	-11.7	32'	88%	136° W	-G	

Mercury and Venus

 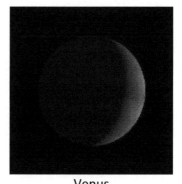

Mercury
5th

Venus
5th

Mercury

Date	Con.	R.A.	Dec.	Mag.	Diam.	Ill.	Elon.	Vis.	Rat.	Close To
1st	Ari	2h 21m	13° 24'	-1.8	5"	99%	4° W	NV	N/A	Uranus
3rd	Ari	2h 38m	15° 3'	-2.1	5"	100%	2° W	NV	N/A	Uranus
5th	Ari	2h 54m	16° 40'	-2.2	5"	100%	1° E	NV	N/A	
7th	Ari	3h 11m	18° 13'	-2.0	5"	99%	3° E	NV	N/A	Pleiades
9th	Ari	3h 29m	19° 39'	-1.8	5"	97%	5° E	NV	N/A	Pleiades

Venus

Date	Con.	R.A.	Dec.	Mag.	Diam.	Ill.	Elon.	Vis.	Rat.	Close To
1st	Tau	5h 11m	27° 47'	-4.5	39"	24%	39° E	PM	****	Aldebaran
3rd	Tau	5h 15m	27° 49'	-4.5	40"	23%	38° E	PM	****	
5th	Tau	5h 18m	27° 48'	-4.5	42"	21%	37° E	PM	****	
7th	Tau	5h 20m	27° 46'	-4.5	43"	19%	35° E	PM	****	
9th	Tau	5h 21m	27° 41'	-4.5	45"	17%	34° E	PM	****	

Mars and the Outer Planets

Mars
5th

Jupiter
5th

Saturn
5th

Mars

Date	Con.	R.A.	Dec.	Mag.	Diam.	Ill.	Elon.	Vis.	Rat.	Close To
1st	Cap	21h 39m	-15° 44'	0.4	8"	86%	74° W	AM	**	
5th	Cap	21h 50m	-14° 54'	0.4	8"	86%	75° W	AM	**	
10th	Aqr	22h 3m	-13° 50'	0.3	8"	86%	77° W	AM	**	

The Outer Planets

Planet	Date	Con.	R.A.	Dec.	Mag.	Diam.	Elon.	Vis.	Rat.	Close To
Jupiter	5th	Sgr	19h 56m	-20° 52'	-2.4	41"	104° W	AM	***	Saturn
Saturn	5th	Cap	20h 16m	-19° 50'	0.6	17"	99° W	AM	***	
Uranus	5th	Ari	2h 19m	13° 28'	5.9	3"	8° W	NV	N/A	Mercury
Neptune	5th	Aqr	23h 26m	-4° 47'	7.9	2"	51° W	AM	**	

Highlights

Date	Time (UT)	Event
2nd	03:49	The waxing gibbous Moon is north of the bright star Regulus. (Leo, evening sky.)
4th	21:28	Mercury is at superior conjunction with the Sun. (Not visible.)
5th	15:39	The waxing gibbous Moon is north of the bright star Spica. (Virgo, evening sky.)
6th	N/A	The Eta Aquariid meteor shower is at its maximum. (ZHR: 70)
7th	10:46	Full Moon. (Visible all night.)
8th	20:32	The just-past full Moon is north of the bright star Antares. (Scorpius, visible all night.)
9th	N/A	The Eta Lyrid meteor shower is at its maximum. (ZHR: 3)

Planet Locations – May 5th

Sun

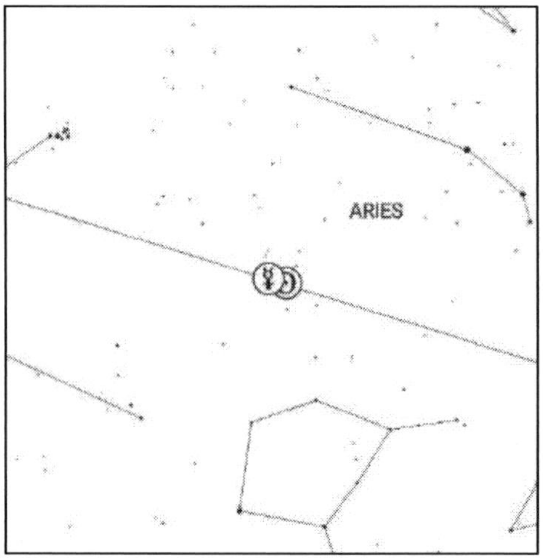

Mercury

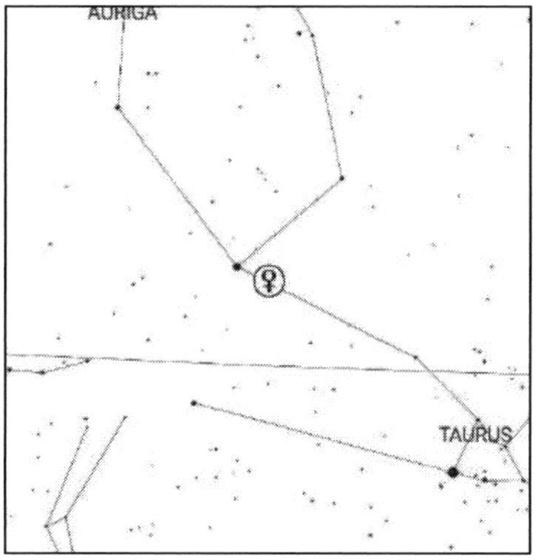

Venus

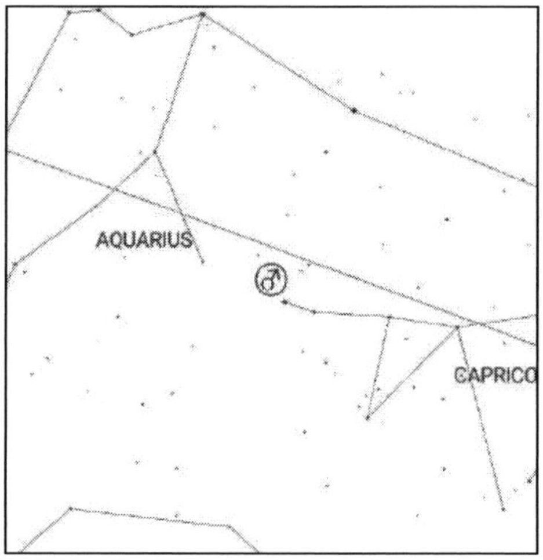

Venus

Mars

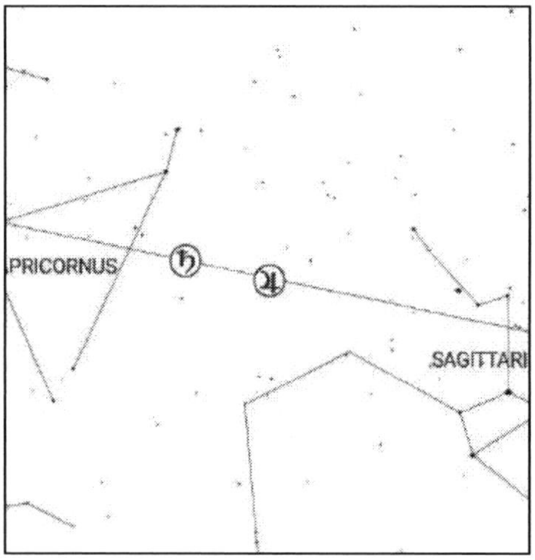

Jupiter

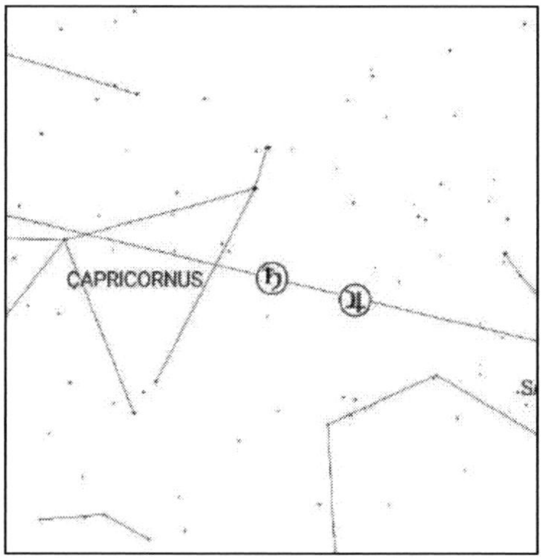

Saturn

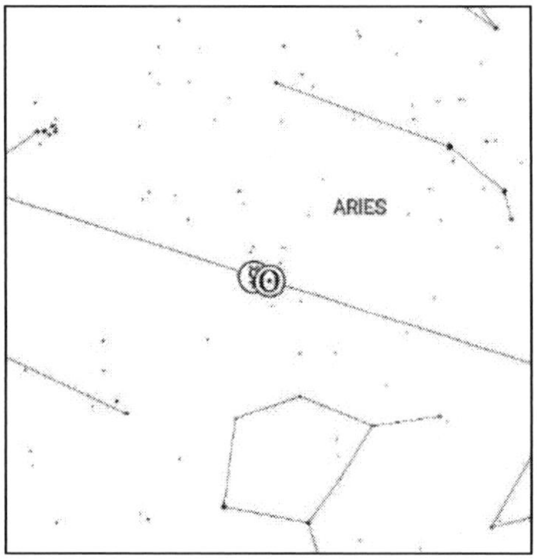

May 1st – Evening Sky

Having reached first quarter yesterday, the Moon is growing fuller and, once again, passes close to Regulus.

Often considered to be a spring constellation, Leo is now beginning to set over the south-western horizon. This means you probably won't see the pair so close again in the evening sky this year.

This image depicts the sky looking south-west at about two hours after sunset.

May 5th – Evening Sky

Four days later and the Moon is passing through another spring constellation, Virgo (Vir) the Virgin. Still a few days away from turning full, the Moon appears to the left of Spica tonight.

Look out too for Arcturus, a red giant star just 36 light years away. It's visible throughout most of the year for observers in the northern hemisphere.

This image depicts the sky looking south-east at about two hours after sunset.

May 5th to 6th – Evening & Morning Sky

Just about two weeks after the Lyrid meteors graced our skies, another meteor shower reaches its maximum – but you'll have to get up early to make the most of it.

The Eta Aquariid meteors appear to originate from the autumn constellation of Aquarius (Aqr) and that constellation doesn't rise until well after midnight. Unfortunately, the Moon is almost full too, so your chances of seeing the theoretical maximum of 70 meteors an hour might be slim!

This image depicts the sky looking south-east about an hour and 45 minutes before sunrise.

May 11th to 20th, 2020

The Moon

   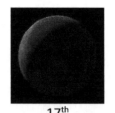 

| 11th | | 13th | | 15th | | 17th | | 19th |

Date	Con	R.A.	Dec	Mag	Diam	Ill.	Elon.	Phase		Close To
11th	Sgr	19h 5m	-23° 58'	-11.3	31'	80%	122° W	-G		
12th	Sgr	20h 2m	-22° 58'	-10.9	31'	70%	109° W	-G		Jupiter, Saturn
13th	Cap	20h 56m	-20° 49'	-10.6	30'	61%	97° W	LQ		Saturn
14th	Cap	21h 47m	-17° 44'	-10.2	30'	51%	85° W	LQ		Mars
15th	Aqr	22h 35m	-13° 57'	-9.7	30'	41%	74° W	LQ		Mars
16th	Aqr	23h 21m	-9° 40'	-9.2	30'	32%	63° W	-Cr		Neptune
17th	Psc	0h 5m	-5° 5'	-8.7	29'	23%	53° W	-Cr		Neptune
18th	Cet	0h 48m	0° 18'	-8.0	29'	16%	44° W	-Cr		
19th	Psc	1h 32m	4° 29'	-7.3	30'	10%	34° W	NM		
20th	Cet	2h 16m	9° 10'	-6.4	30'	5%	24° W	NM		Uranus

Mercury and Venus

Mercury
15th

Venus
15th

Mercury

Date	Con.	R.A.	Dec.	Mag.	Diam.	Ill.	Elon.	Vis.	Rat.	Close To
11th	Tau	3h 46m	20° 58'	-1.6	5"	95%	8° E	NV	N/A	Pleiades
13th	Tau	4h 4m	22° 8'	-1.4	5"	91%	10° E	NV	N/A	Pleiades, Hyades, Aldebara
15th	Tau	4h 21m	23° 8'	-1.2	6"	86%	13° E	NV	N/A	Pleiades, Hyades, Aldebara
17th	Tau	4h 38m	23° 58'	-1.0	6"	81%	15° E	NV	N/A	Hyades, Aldebaran
19th	Tau	4h 54m	24° 37'	-0.8	6"	76%	17° E	PM	**	Venus, Hyades, Aldebaran

Venus

Date	Con.	R.A.	Dec.	Mag.	Diam.	Ill.	Elon.	Vis.	Rat.	Close To
11th	Tau	5h 22m	27° 33'	-4.4	46"	15%	32° E	PM	***	
13th	Tau	5h 23m	27° 24'	-4.4	47"	13%	30° E	PM	***	
15th	Tau	5h 22m	27° 12'	-4.4	49"	11%	28° E	PM	***	
17th	Tau	5h 21m	26° 57'	-4.3	50"	9%	26° E	PM	***	
19th	Tau	5h 20m	26° 39'	-4.3	52"	8%	23° E	PM	***	Mercury

Mars and the Outer Planets

Mars	Jupiter	Saturn
15th	15th	15th

Mars

Date	Con.	R.A.	Dec.	Mag.	Diam.	Ill.	Elon.	Vis.	Rat.	Close To
11th	Aqr	22h 6m	-13° 37'	0.3	8"	86%	77° W	AM	**	
15th	Aqr	22h 17m	-12° 45'	0.2	8"	85%	78° W	AM	**	Moon
20th	Aqr	22h 30m	-11° 38'	0.1	9"	85%	80° W	AM	**	

The Outer Planets

Planet	Date	Con.	R.A.	Dec.	Mag.	Diam.	Elon.	Vis.	Rat.	Close To
Jupiter	15th	Sgr	19h 57m	-20° 52'	-2.5	43"	113° W	AM	****	Saturn
Saturn	15th	Cap	20h 16m	-19° 51'	0.5	17"	109° W	AM	***	
Uranus	15th	Ari	2h 21m	13° 39'	5.9	3"	17° W	AM	*	
Neptune	15th	Aqr	23h 26m	-4° 42'	7.9	2"	61° W	AM	**	

Highlights

Date	Time (UT)	Event
11th	07:33	Saturn is stationary prior to beginning retrograde motion. (Morning sky.)
12th	05:43	The waning gibbous Moon is south of dwarf planet Pluto. (Morning sky.)
	10:58	The waning gibbous Moon is south of Jupiter. (Morning sky.)
	17:53	The waning gibbous Moon is south of Saturn. (Morning sky.)
13th	08:20	Venus is stationary prior to beginning retrograde motion. (Evening sky.)
14th	14:03	Last Quarter Moon. (Morning sky.)
	17:44	Jupiter is stationary prior to beginning retrograde motion. (Morning sky.)
15th	00:40	The just-past last quarter Moon is south of Mars. (Morning sky.)
16th	16:13	The waning crescent Moon is south of Neptune. (Morning sky.)
19th	N/A	Good opportunity to see Earthshine on the waning crescent Moon. (Morning sky.)
20th	16:58	The waning crescent Moon is south of Uranus. (Morning sky.)

Planet Locations – May 15th

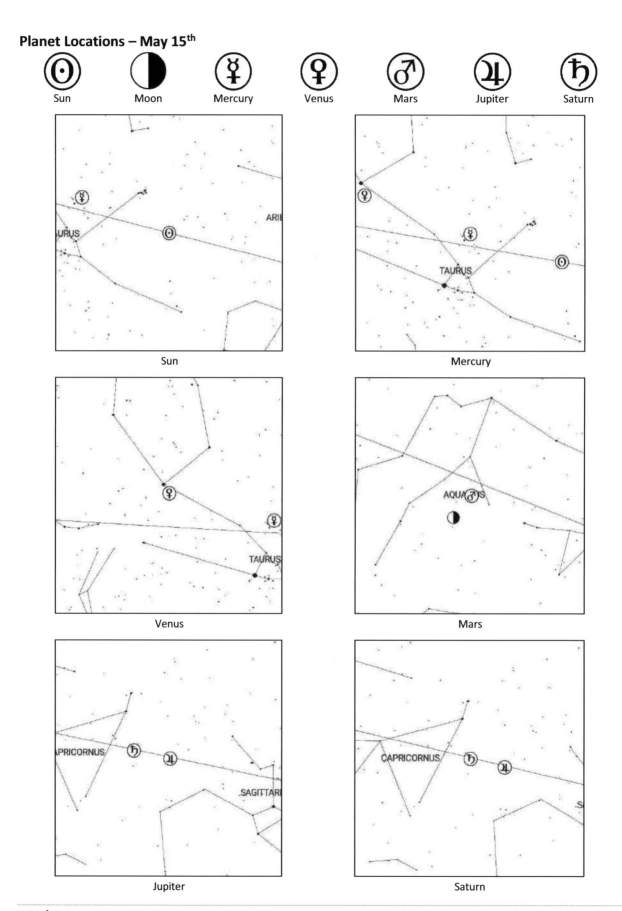

Sun   Moon   Mercury   Venus   Mars   Jupiter   Saturn

Sun

Mercury

Venus

Mars

Jupiter

Saturn

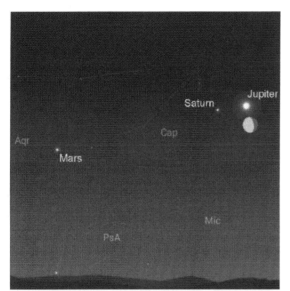

May 12th – Morning Sky

The Moon is approaching last quarter and has returned to the planets in the pre-dawn sky. This morning it appears closest to Jupiter and hovers almost directly below the giant planet.

Meanwhile, Saturn is a little way to the left while Mars awaits toward the east.

This image depicts the sky looking south-east at about an hour before sunrise.

May 15th – Morning Sky

Three days later and the Moon, after reaching last quarter yesterday, is now to the lower left of Mars.

This is the last lunar encounter of the morning sky before the Moon returns to the evening twilight in about 8 or 9 days time.

This image depicts the sky looking south-east at about an hour before sunrise.

May 20th – Evening Sky

There's a great opportunity to see Mercury after sunset tonight – especially if you've never seen the tiny planet before.

It's beginning to emerge from out of the Sun's glare and may be glimpsed just to the lower right of Venus. The pair will easily fit within the same binocular field of view and will be at their closest tomorrow.

This image depicts the sky looking north-west at about 30 minutes after sunset.

May 21st to 31st, 2020

The Moon

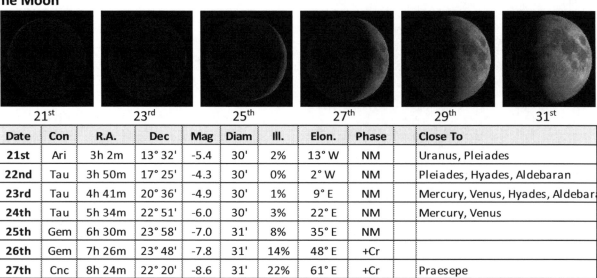

| 21st | 23rd | 25th | 27th | 29th | 31st |

Date	Con	R.A.	Dec	Mag	Diam	Ill.	Elon.	Phase	Close To
21st	Ari	3h 2m	13° 32'	-5.4	30'	2%	13° W	NM	Uranus, Pleiades
22nd	Tau	3h 50m	17° 25'	-4.3	30'	0%	2° W	NM	Pleiades, Hyades, Aldebaran
23rd	Tau	4h 41m	20° 36'	-4.9	30'	1%	9° E	NM	Mercury, Venus, Hyades, Aldebar
24th	Tau	5h 34m	22° 51'	-6.0	30'	3%	22° E	NM	Mercury, Venus
25th	Gem	6h 30m	23° 58'	-7.0	31'	8%	35° E	NM	
26th	Gem	7h 26m	23° 48'	-7.8	31'	14%	48° E	+Cr	
27th	Cnc	8h 24m	22° 20'	-8.6	31'	22%	61° E	+Cr	Praesepe
28th	Cnc	9h 20m	19° 35'	-9.2	32'	32%	74° E	+Cr	Praesepe, Regulus
29th	Leo	10h 15m	15° 42'	-9.8	32'	43%	87° E	FQ	Regulus
30th	Leo	11h 9m	10° 54'	-10.3	32'	54%	100° E	FQ	
31st	Vir	12h 2m	5° 27'	-10.7	32'	65%	112° E	FQ	

Mercury and Venus

Mercury
25th

Venus
25th

Mercury

Date	Con.	R.A.	Dec.	Mag.	Diam.	Ill.	Elon.	Vis.	Rat.	Close To
21st	Tau	5h 10m	25° 6'	-0.6	6"	71%	19° E	PM	***	Venus, Aldebaran
23rd	Tau	5h 25m	25° 26'	-0.5	6"	65%	20° E	PM	***	Moon, Venus
25th	Tau	5h 39m	25° 37'	-0.3	7"	60%	22° E	PM	***	Venus
27th	Tau	5h 52m	25° 40'	-0.1	7"	55%	23° E	PM	***	
29th	Gem	6h 4m	25° 35'	0.0	7"	50%	24° E	PM	***	
31st	Gem	6h 15m	25° 24'	0.2	7"	46%	25° E	PM	***	

Venus

Date	Con.	R.A.	Dec.	Mag.	Diam.	Ill.	Elon.	Vis.	Rat.	Close To
21st	Tau	5h 17m	26° 19'	-4.2	53"	6%	21° E	PM	***	Mercury
23rd	Tau	5h 14m	25° 55'	-4.2	54"	4%	18° E	PM	***	Moon, Mercury, Aldebaran
25th	Tau	5h 10m	25° 29'	-4.1	55"	3%	15° E	NV	N/A	Mercury, Aldebaran
27th	Tau	5h 6m	24° 60'	-4.0	56"	2%	12° E	NV	N/A	Hyades, Aldebaran
29th	Tau	5h 1m	24° 28'	-3.9	57"	1%	9° E	NV	N/A	Hyades, Aldebaran
31st	Tau	4h 56m	23° 54'	-3.8	58"	0%	5° E	NV	N/A	Hyades, Aldebaran

Mars and the Outer Planets

Mars
25th

Jupiter
25th

Saturn
25th

Mars

Date	Con.	R.A.	Dec.	Mag.	Diam.	Ill.	Elon.	Vis.	Rat.	Close To
21st	Aqr	22h 33m	-11° 24'	0.1	9"	85%	80° W	AM	**	
25th	Aqr	22h 43m	-10° 29'	0.1	9"	85%	82° W	AM	**	
31st	Aqr	22h 58m	-9° 6'	0.0	9"	85%	84° W	AM	**	

The Outer Planets

Planet	Date	Con.	R.A.	Dec.	Mag.	Diam.	Elon.	Vis.	Rat.	Close To
Jupiter	25th	Sgr	19h 56m	-20° 56'	-2.5	44"	124° W	AM	****	Saturn
Saturn	25th	Cap	20h 16m	-19° 54'	0.5	18"	119° W	AM	***	
Uranus	25th	Ari	2h 24m	13° 49'	5.9	3"	27° W	AM	*	
Neptune	25th	Aqr	23h 27m	-4° 38'	7.9	2"	71° W	AM	**	

Highlights

Date	Time (UT)	Event
21st	07:49	Mercury is 0.9° south of Venus. (Evening sky.)
22nd	09:45	New Moon. (Not visble.)
24th	02:19	The waxing crescent Moon is south of Venus. (Evening sky.)
	09:43	The waxing crescent Moon is south of Mercury. (Evening sky.)
	N/A	The May Camelopardalid meteor shower is at its maximum. (ZHR: Variable.)
25th	N/A	Good opportunity to see Earthshine on the waxing crescent Moon. (Evening sky.)
27th	19:38	The waxing crescent Moon is north of the Praesepe star cluster. (Cancer, evening sky.)
29th	08:00	The nearly first quarter Moon is north of the bright star Regulus. (Leo, evening sky.)
30th	03:30	First Quarter Moon. (Evening sky.)

Planet Locations – May 25th

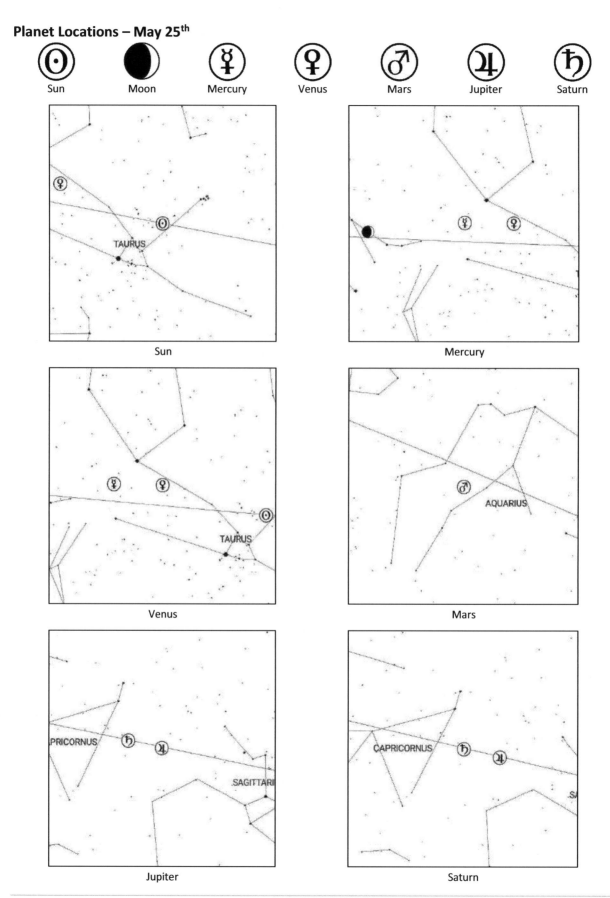

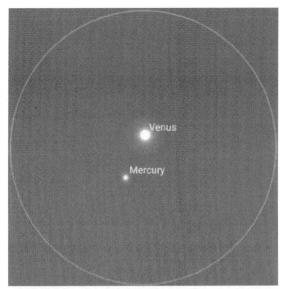

May 21st – Evening Sky

Mercury continues to distance itself from the Sun and climb higher into the evening twilight. Venus, meanwhile, is sinking toward the Sun, so the two appear to be passing one another.

They're at their closest tonight, with roughly a degree between them. Keep an eye on them to see how they move over the next few nights.

This image depicts the view through 10x50 binoculars, looking north-west about 30 minutes after sunset.

May 24th – Evening Sky

If you've been watching Mercury and Venus over the past few evenings, you'll have seen the pair pass each other and grow farther apart.

You may have also glimpsed the thin crescent Moon last night, but tonight it's more easily seen to the upper left of the two. All three should appear almost equally spaced, with Mercury in the middle.

This image depicts the sky looking north-west at about 30 minutes after sunset.

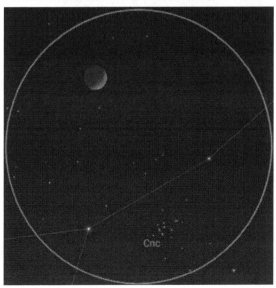

May 27th – Evening Sky

There's a good opportunity to catch the crescent Moon close to an open cluster tonight. The Praesepe, or Beehive Cluster, can be found at the heart of Cancer (Cnc) the Crab. (It's the little group of stars close to the letters Cnc in the image to the left.)

Grab your binoculars and take a look at the Moon – the cluster should just be visible to the lower right.

This image depicts the view through 10x50 binoculars looking west about two hours after sunset.

June 1st to 10th, 2020

The Moon

| | 1st | | 3rd | | 5th | | 7th | | 9th |

Date	Con	R.A.	Dec	Mag	Diam	Ill.	Elon.	Phase		Close To
1st	Vir	12h 55m	0° 23'	-11.2	33'	76%	124° E	+G		Spica
2nd	Vir	13h 48m	-6° 16'	-11.5	33'	85%	136° E	+G		Spica
3rd	Lib	14h 43m	-11° 49'	-11.9	33'	93%	149° E	+G		
4th	Lib	15h 40m	-16° 42'	-12.3	33'	98%	162° E	FM		
5th	Oph	16h 39m	-20° 32'	-12.6	32'	100%	176° E	FM		Antares
6th	Oph	17h 40m	-23° 2'	-12.5	32'	99%	170° W	FM		
7th	Sgr	18h 41m	-24° 3'	-12.2	32'	96%	156° W	FM		
8th	Sgr	19h 40m	-23° 36'	-11.8	31'	91%	142° W	-G		Jupiter, Saturn
9th	Cap	20h 36m	-21° 50'	-11.5	31'	84%	129° W	-G		Jupiter, Saturn
10th	Cap	21h 30m	-19° 1'	-11.2	30'	76%	117° W	-G		

Mercury and Venus

Mercury
5th

Venus
5th

Mercury

Date	Con.	R.A.	Dec.	Mag.	Diam.	Ill.	Elon.	Vis.	Rat.	Close To
1st	Gem	6h 20m	25° 16'	0.3	8"	44%	25° E	PM	****	
3rd	Gem	6h 30m	24° 57'	0.5	8"	39%	26° E	PM	****	
5th	Gem	6h 38m	24° 34'	0.6	8"	35%	26° E	PM	****	
7th	Gem	6h 46m	24° 7'	0.8	9"	31%	25° E	PM	****	
9th	Gem	6h 52m	23° 37'	1.0	9"	27%	25° E	PM	****	

Venus

Date	Con.	R.A.	Dec.	Mag.	Diam.	Ill.	Elon.	Vis.	Rat.	Close To
1st	Tau	4h 54m	23° 36'	-3.8	58"	0%	4° E	NV	N/A	Hyades, Aldebaran
3rd	Tau	4h 49m	22° 59'	-3.7	58"	0%	0° E	NV	N/A	Hyades, Aldebaran
5th	Tau	4h 44m	22° 22'	-3.8	58"	0%	3° W	NV	N/A	Hyades, Aldebaran
7th	Tau	4h 39m	21° 44'	-3.9	57"	1%	6° W	NV	N/A	Hyades, Aldebaran
9th	Tau	4h 34m	21° 7'	-4.0	57"	1%	10° W	NV	N/A	Hyades, Aldebaran

Mars and the Outer Planets

Mars
5th

Jupiter
5th

Saturn
5th

Mars

Date	Con.	R.A.	Dec.	Mag.	Diam.	Ill.	Elon.	Vis.	Rat.	Close To
1st	Aqr	23h 1m	-8° 52'	0.0	9"	85%	85° W	AM	**	
5th	Aqr	23h 11m	-7° 57'	-0.1	10"	85%	86° W	AM	**	Neptune
10th	Aqr	23h 23m	-6° 47'	-0.2	10"	84%	88° W	AM	**	Neptune

The Outer Planets

Planet	Date	Con.	R.A.	Dec.	Mag.	Diam.	Elon.	Vis.	Rat.	Close To
Jupiter	5th	Sgr	19h 54m	-21° 4'	-2.6	45"	135° W	AM	****	Saturn
Saturn	5th	Cap	20h 14m	-19° 59'	0.4	18"	130° W	AM	****	
Uranus	5th	Ari	2h 26m	14° 0'	5.9	3"	37° W	AM	*	
Neptune	5th	Aqr	23h 28m	-4° 35'	7.9	2"	82° W	AM	***	Mars

Highlights

Date	Time (UT)	Event
1st	02:44	The waxing gibbous Moon is north of the bright star Spica. (Virgo, evening sky.)
3rd	17:38	Venus is at inferior conjunction with the Sun. (Not visible.)
4th	13:04	Mercury is at greatest eastern elongation from the Sun. (Evening sky.)
5th	02:48	The almost full Moon is north of the bright star Antares. (Scorpius, visible all night.)
	19:13	Full Moon. (Visible all night.)
	19:24	Penumbral lunar eclipse. Visible from Africa, Antarctica, Asia, the eastern Atlantic, Australia, Europe, the Indian Ocean and eastern South America.
8th	14:12	The waning gibbous Moon is south of dwarf planet Pluto. (Morning sky.)
	16:44	The waning gibbous Moon is south of Jupiter. (Morning sky.)
9th	01:50	The waning gibbous Moon is south of Saturn. (Morning sky.)

Planet Locations – June 5th

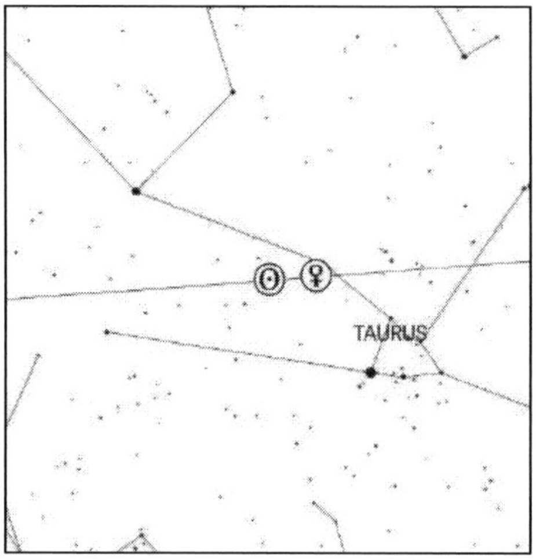

Sun	Mercury	Venus	Mars	Jupiter	Saturn

Sun

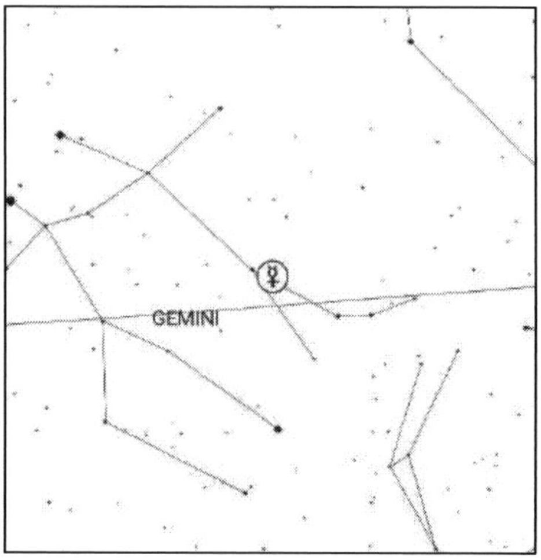

Mercury

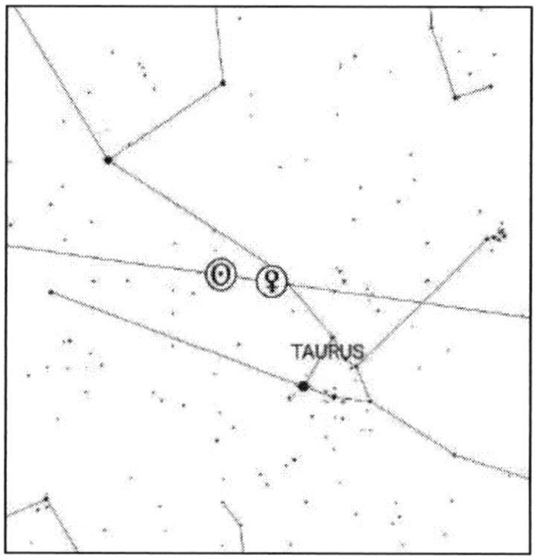

Venus

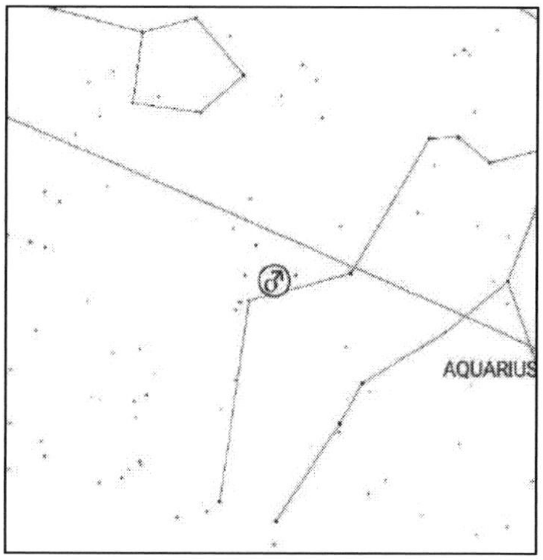

Mars

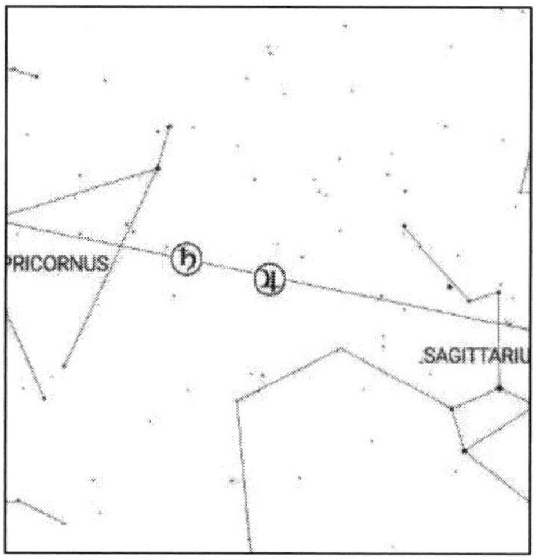

Jupiter

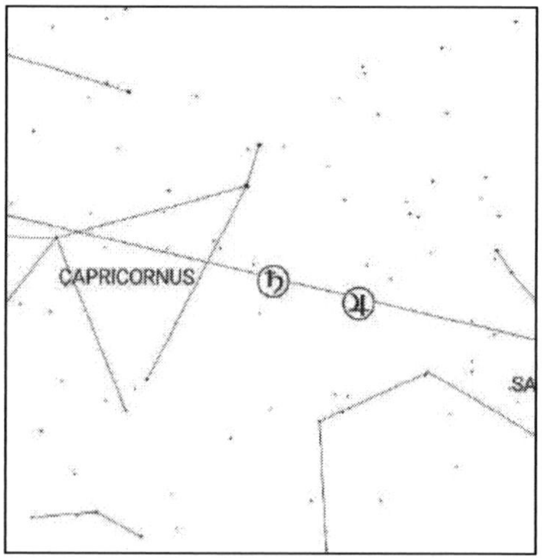

Saturn

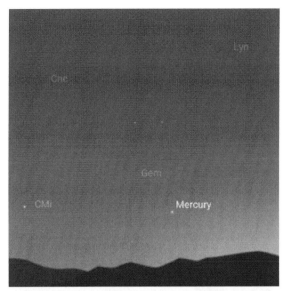

June 4th – Evening Sky

Mercury is now at its furthest distance from the Sun in the sky, but unfortunately, there's no bright Moon nearby and Venus has vanished from view.

That being said, you should still be able to glimpse it as a pinkish-white "star" in the evening twilight, just above the western horizon.

This image depicts the sky looking west at about 30 minutes after sunset.

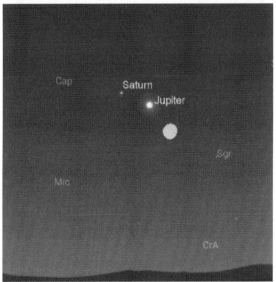

June 8th – Morning Sky

The Moon turned full three days ago and is now close again to Jupiter and Saturn. You can either wait until just before midnight on the 7th to see the trio rising together, or else you can rise early on the 8th.

If you're undecided, you'll get a better view if you choose to rise early on the 8th, as the three worlds will be higher above the horizon.

This image depicts the sky looking south at about 45 minutes before sunrise.

June 9th – Morning Sky

Yesterday the waning gibbous Moon appeared to the right of Jupiter and Saturn. This morning you'll see our closest celestial companion has skipped past them both and is now to their left.

It'll catch up to Mars in another three days' time.

This image depicts the sky looking south at about 45 minutes before sunrise.

June 11th to 20th, 2020

The Moon

11th

13th

15th

17th

19th

Date	Con	R.A.	Dec	Mag	Diam	Ill.	Elon.	Phase		Close To
11th	Aqr	22h 20m	-15° 23'	-10.8	30'	67%	105° W	-G		
12th	Aqr	23h 7m	-11° 11'	-10.4	30'	57%	94° W	LQ		Mars, Neptune
13th	Aqr	23h 51m	-6° 37'	-10.0	30'	48%	84° W	LQ		Mars, Neptune
14th	Cet	0h 35m	-1° 52'	-9.6	30'	38%	74° W	LQ		
15th	Psc	1h 18m	2° 57'	-9.1	30'	29%	65° W	-Cr		
16th	Psc	2h 2m	7° 41'	-8.5	30'	21%	55° W	-Cr		Uranus
17th	Ari	2h 47m	12° 11'	-7.8	30'	14%	45° W	-Cr		Uranus
18th	Tau	3h 35m	16° 15'	-7.0	30'	8%	34° W	NM		Venus, Pleiades
19th	Tau	4h 25m	19° 42'	-6.0	30'	3%	22° W	NM		Venus, Pleiades, Hyades, Aldebar
20th	Tau	5h 18m	22° 17'	-4.9	31'	1%	10° W	NM		Aldebaran

Mercury and Venus

Mercury
15th

Venus
15th

Mercury

Date	Con.	R.A.	Dec.	Mag.	Diam.	Ill.	Elon.	Vis.	Rat.	Close To
11th	Gem	6h 56m	23° 5'	1.3	10"	24%	24° E	PM	****	
13th	Gem	7h 0m	22° 32'	1.5	10"	20%	23° E	PM	***	
15th	Gem	7h 2m	21° 59'	1.8	10"	17%	21° E	PM	***	
17th	Gem	7h 3m	21° 26'	2.1	11"	13%	19° E	PM	***	
19th	Gem	7h 3m	20° 53'	2.5	11"	10%	17° E	PM	***	

Venus

Date	Con.	R.A.	Dec.	Mag.	Diam.	Ill.	Elon.	Vis.	Rat.	Close To
11th	Tau	4h 30m	20° 31'	-4.0	56"	2%	13° W	NV	N/A	Hyades, Aldebaran
13th	Tau	4h 26m	19° 57'	-4.1	55"	3%	16° W	AM	**	Pleiades, Hyades, Aldebaran
15th	Tau	4h 23m	19° 25'	-4.2	54"	5%	19° W	AM	***	Pleiades, Hyades, Aldebaran
17th	Tau	4h 20m	18° 57'	-4.2	53"	6%	21° W	AM	***	Pleiades, Hyades, Aldebaran
19th	Tau	4h 18m	18° 32'	-4.3	51"	8%	24° W	AM	***	Moon, Pleiades, Hyades, Aldebar:

Mars and the Outer Planets

Mars
15th

Jupiter
15th

Saturn
15th

Mars

Date	Con.	R.A.	Dec.	Mag.	Diam.	Ill.	Elon.	Vis.	Rat.	Close To
11th	Aqr	23h 25m	-6° 33'	-0.2	10"	84%	89° W	AM	**	Neptune
15th	Aqr	23h 35m	-5° 38'	-0.2	10"	84%	90° W	AM	**	Neptune
20th	Aqr	23h 47m	-4° 29'	-0.3	11"	84%	93° W	AM	***	Neptune

The Outer Planets

Planet	Date	Con.	R.A.	Dec.	Mag.	Diam.	Elon.	Vis.	Rat.	Close To
Jupiter	15th	Sgr	19h 50m	-21° 15'	-2.7	46"	147° W	AM	****	Saturn
Saturn	15th	Cap	20h 13m	-20° 6'	0.3	18"	141° W	AM	****	
Uranus	15th	Ari	2h 28m	14° 9'	5.9	3"	47° W	AM	*	
Neptune	15th	Aqr	23h 28m	-4° 34'	7.9	2"	92° W	AM	***	Mars

Highlights

Date	Time (UT)	Event
12th	12:20	Mars is 1.7° south of Neptune. (Morning sky.)
	22:08	The nearly last quarter Moon is south of Neptune. (Morning sky.)
	22:36	The nearly last quarter Moon is south of Mars. (Morning sky.)
13th	06;24	Last Quarter Moon. (Morning sky.)
17th	00:36	The waning crescent Moon is south of Uranus. (Morning sky.)
	19:30	Mercury is stationary before beginning retrograde motion. (Evening sky.)
	N/A	Good opportunity to see Earthshine on the waning crescent Moon. (Morning sky.)
18th	18:47	The waning crescent Moon is south of the Pleiades star cluster. (Taurus, morning sky.)
19th	08:12	The waning crescent Moon is north of Venus. (Morning sky.)
	18:08	The waning crescent Moon is north of the bright star Aldebaran. (Taurus, morning sky.)
20th	21:44	Summer solstice.

Planet Locations – June 15th

| Sun | Mercury | Venus | Mars | Jupiter | Saturn |

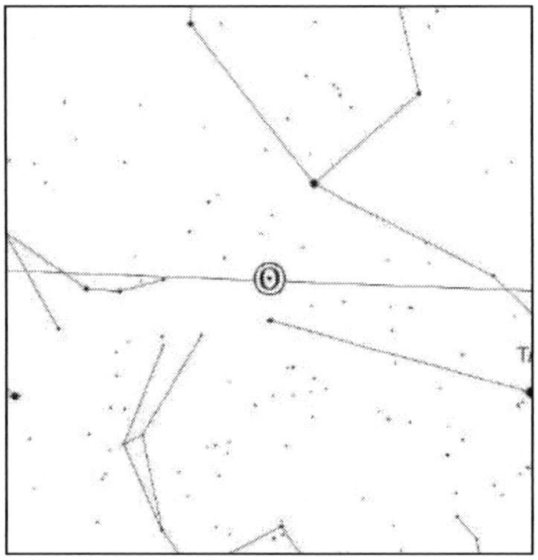

Sun

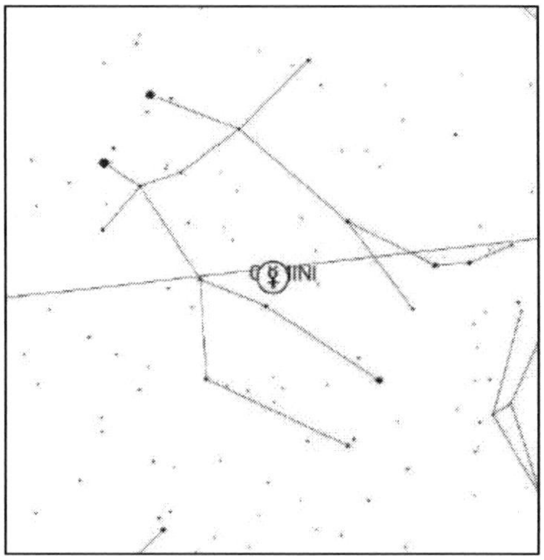

Mercury

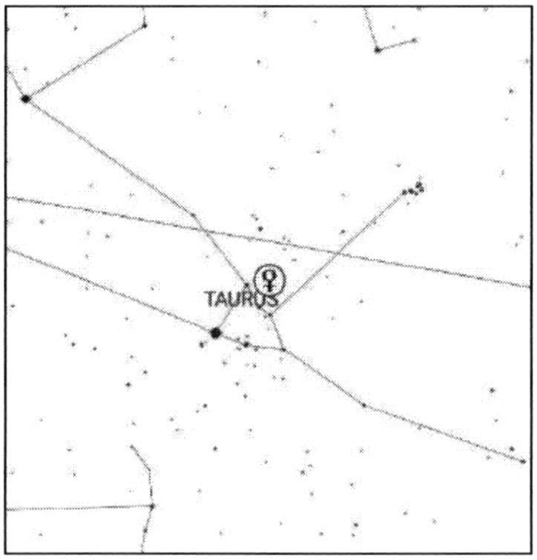

Venus

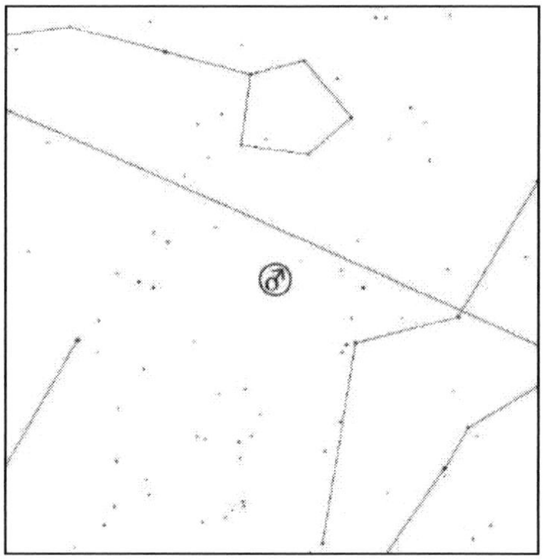

Mars

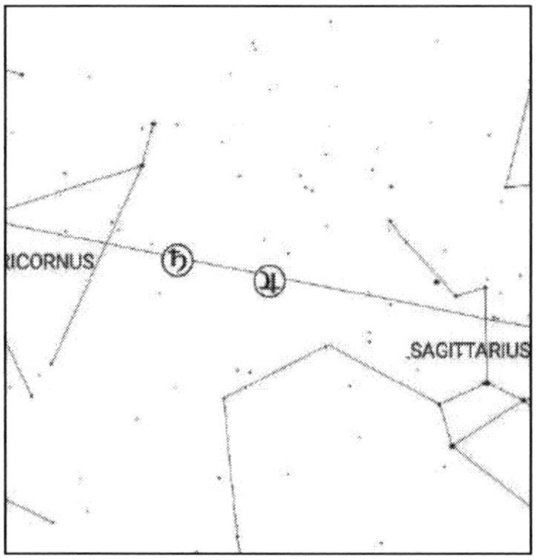

Jupiter

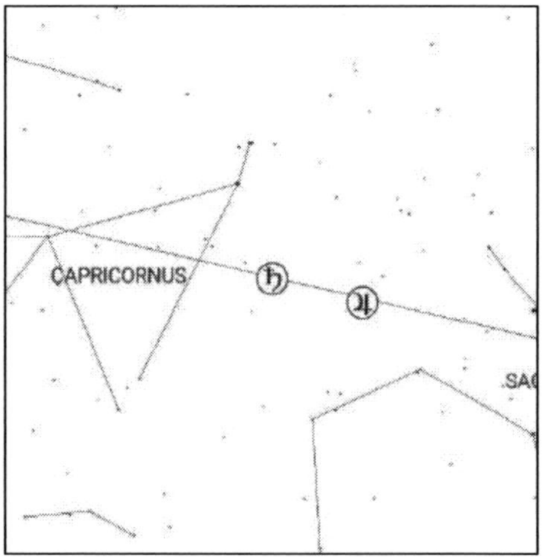

Saturn

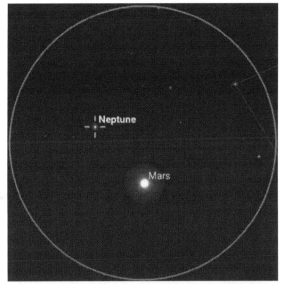

June 12th – Morning Sky

This year has been great for planetary conjunctions and this month continues the trend. Mars and Neptune are less than two degrees apart this morning and can easily fit within the same binocular field of view.

Clouded out? Not to worry; neither planet moves as fast as Mercury or Venus so the pair will be visible together for about another ten days.

This image depicts the sky looking south-east about 90 minutes before sunrise.

June 12th – Morning Sky

Mars has another companion in the pre-dawn sky, this time in the form of the Moon. It's almost reached last quarter and can be found hanging over the southern horizon at sunrise.

If you want to see it close to Mars, you'll need to get up a little earlier, while it's still dark enough to see the brighter stars (and Mars) in the sky.

This image depicts the sky looking south-east at about 45 minutes before sunrise.

June 19th – Morning Sky

Venus is currently edging away from the Sun and is rising earlier in the pre-dawn sky. This morning it's joined by a super slim Moon, just a few days away from turning new.

They're quite low over the horizon but if you have an unobstructed view you should have no problem spotting the pair.

This image depicts the sky looking east at about 45 minutes before sunrise.

June 21st to 30th, 2020

The Moon

| 21st | 23rd | 25th | 27th | 29th |

Date	Con	R.A.	Dec	Mag	Diam	Ill.	Elon.	Phase	Close To
21st	Gem	6h 14m	23° 47'	-4.2	31'	0%	3° E	NM	Mercury
22nd	Gem	7h 11m	23° 60'	-5.5	31'	2%	16° E	NM	Mercury
23rd	Cnc	8h 9m	22° 51'	-6.6	31'	6%	30° E	NM	Praesepe
24th	Cnc	9h 7m	20° 22'	-7.6	32'	12%	43° E	NM	Praesepe
25th	Leo	10h 3m	16° 42'	-8.4	32'	20%	56° E	+Cr	Regulus
26th	Leo	10h 57m	12° 4'	-9.1	32'	29%	69° E	+Cr	
27th	Vir	11h 50m	6° 46'	-9.7	32'	40%	81° E	FQ	
28th	Vir	12h 42m	1° 5'	-10.2	32'	52%	93° E	FQ	Spica
29th	Vir	13h 34m	-4° 41'	-10.7	32'	63%	105° E	FQ	Spica
30th	Lib	14h 27m	-10° 14'	-11.1	32'	74%	117° E	+G	

Mercury and Venus

Mercury
25th

Venus
25th

Mercury

Date	Con.	R.A.	Dec.	Mag.	Diam.	Ill.	Elon.	Vis.	Rat.	Close To
21st	Gem	7h 1m	20° 22'	2.9	11"	7%	15° E	NV	N/A	Moon
23rd	Gem	6h 58m	19° 54'	3.4	12"	5%	12° E	NV	N/A	
25th	Gem	6h 54m	19° 29'	3.9	12"	3%	9° E	NV	N/A	
27th	Gem	6h 50m	19° 7'	4.5	12"	2%	6° E	NV	N/A	
29th	Gem	6h 45m	18° 49'	4.9	12"	1%	2° E	NV	N/A	

Venus

Date	Con.	R.A.	Dec.	Mag.	Diam.	Ill.	Elon.	Vis.	Rat.	Close To
21st	Tau	4h 17m	18° 10'	-4.3	50"	10%	26° W	AM	***	Pleiades, Hyades, Aldebaran
23rd	Tau	4h 16m	17° 53'	-4.4	48"	12%	29° W	AM	***	Pleiades, Hyades, Aldebaran
25th	Tau	4h 16m	17° 38'	-4.4	47"	13%	31° W	AM	***	Pleiades, Hyades, Aldebaran
27th	Tau	4h 16m	17° 27'	-4.4	46"	15%	33° W	AM	***	Pleiades, Hyades, Aldebaran
29th	Tau	4h 17m	17° 19'	-4.4	44"	17%	34° W	AM	****	Pleiades, Hyades, Aldebaran

Mars and the Outer Planets

Mars
25th

Jupiter
25th

Saturn
25th

Mars

Date	Con.	R.A.	Dec.	Mag.	Diam.	Ill.	Elon.	Vis.	Rat.	Close To
21st	Aqr	23h 49m	-4° 16'	-0.3	11"	84%	93° W	AM	***	
25th	Psc	23h 59m	-3° 22'	-0.4	11"	84%	95° W	AM	***	
30th	Psc	0h 10m	-2° 17'	-0.5	11"	84%	97° W	AM	***	

The Outer Planets

Planet	Date	Con.	R.A.	Dec.	Mag.	Diam.	Elon.	Vis.	Rat.	Close To
Jupiter	25th	Sgr	19h 46m	-21° 28'	-2.7	47"	158° W	AM	*****	Saturn
Saturn	25th	Cap	20h 10m	-20° 15'	0.3	18"	152° W	AM	****	
Uranus	25th	Ari	2h 29m	14° 17'	5.8	3"	57° W	AM	*	
Neptune	25th	Aqr	23h 28m	-4° 34'	7.9	2"	103° W	AM	***	Mars

Highlights

Date	Time (UT)	Event
21st	06:41	Annular solar eclipse. Visible from eastern Africa, Asia and the Indian Ocean.
	06:42	New Moon. (Not visible.)
23rd	12:58	Neptune is stationary prior to beginning retrograde motion. (Morning sky.)
	21:54	Venus is stationary prior to resuming prograde motion. (Morning sky.)
24th	00:58	The waxing crescent Moon is north of the Praesepe star cluster. (Cancer, evening sky.)
	N/A	Good opportunity to see Earthshine on the waxing crescent Moon. (Evening sky.)
25th	13:20	The waxing crescent Moon is north of the bright star Regulus. (Leo, evening sky.)
27th	N/A	The Bootid meteor shower is at its maximum. (ZHR: Variable.)
28th	08:16	First Quarter Moon. (Evening sky.)
29th	07:33	The just-past first quarter Moon is north of the bright star Spica. (Virgo, evening sky.)

Planet Locations – June 25th

Sun   Mercury   Venus   Mars   Jupiter   Saturn

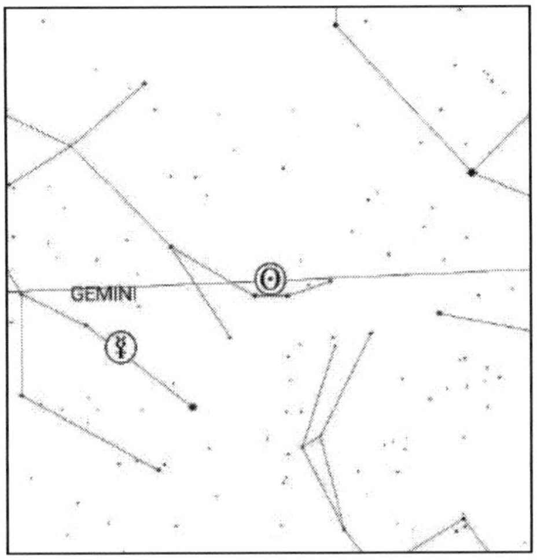

Sun

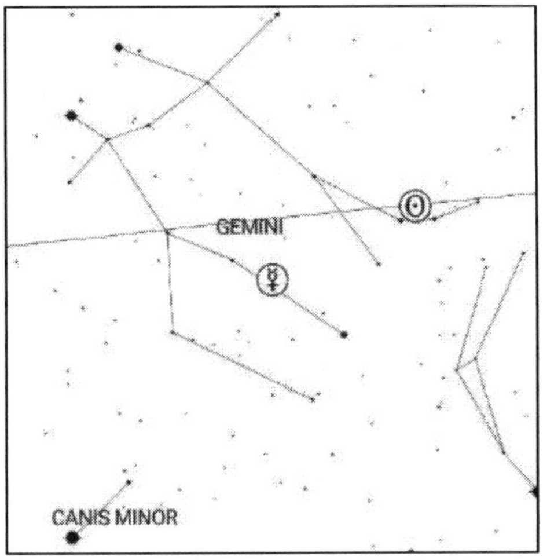

Mercury

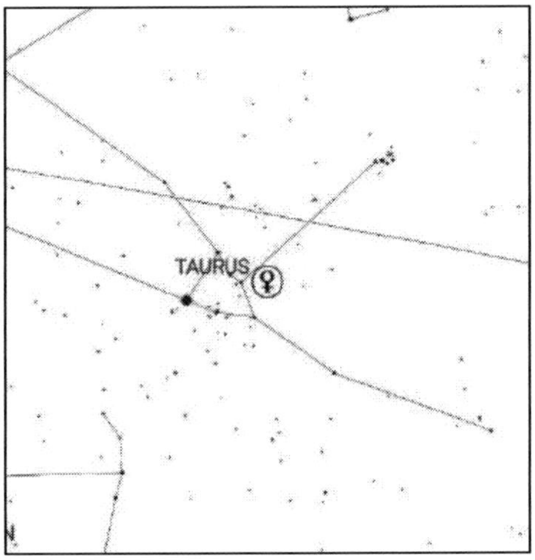

Venus

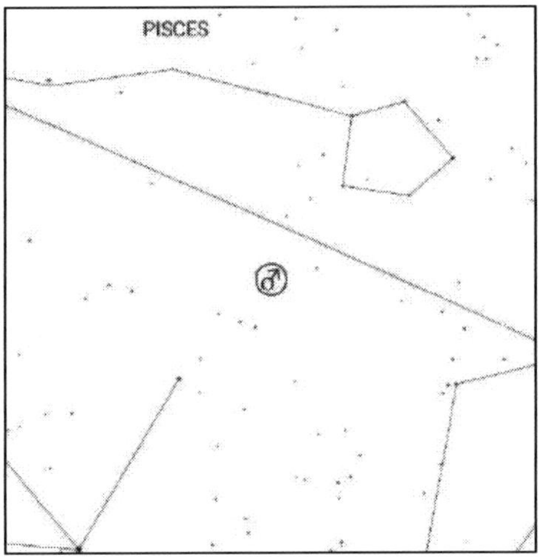

Mars

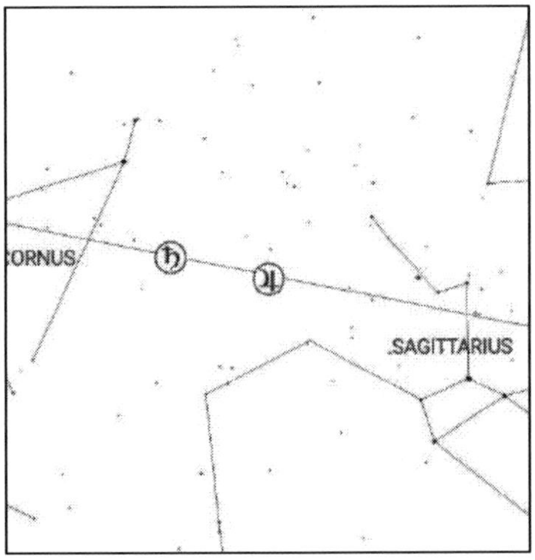

Jupiter

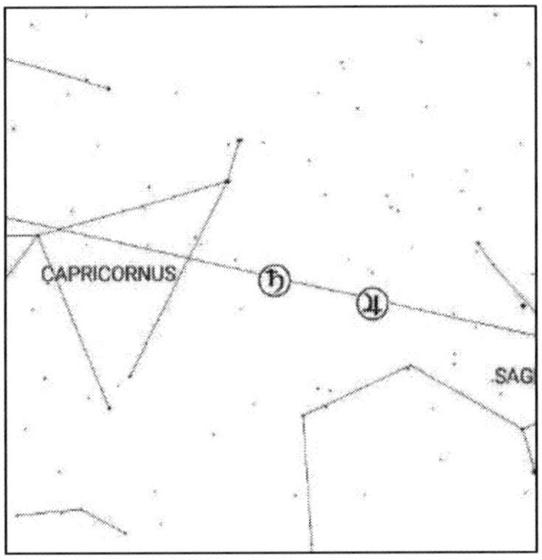

Saturn

June 24th – Evening Sky

The summer solstice occurred on the 20th, which means the nights are currently at their shortest. The Moon turned new on the 21st and is coming around again for another return to the evening sky.

Unfortunately, there are no nearby planets tonight but there's a good opportunity to see Earthshine. Also keep an eye out for Regulus to the upper left.

This image depicts the sky looking west about an hour after sunset.

June 25th – Evening Sky

The summer provides us with a wonderful opportunity to see the Milky Way, our own galaxy, with just your eyes.

You'll find the flowing between Scorpius (Sco) the Scorpion and Sagittarius (Sgr), the Archer. Look toward the X in the image to the left and you'll be staring at the center of our galaxy, some 25,000 light years away!

This image depicts the sky looking south about two hours after sunset.

June 28th – Evening Sky

Four days later and the first quarter Moon has an encounter with Spica, the brightest star in the constellation of Virgo (Vir), the Virgin.

Now that summer has officially started, the stars of spring are beginning to sink into the west in the hours after sunset. We'll have another three months or so before Spica eventually disappears from view.

This image depicts the sky looking south-west about two hours after sunset.

July 1st to 10th, 2020

The Moon

| 1st | 3rd | 5th | 7th | 9th |

Date	Con	R.A.	Dec	Mag	Diam	Ill.	Elon.	Phase	Close To
1st	Lib	15h 22m	-15° 13'	-11.5	32'	83%	130° E	+G	
2nd	Sco	16h 19m	-19° 20'	-11.8	32'	91%	143° E	+G	Antares
3rd	Oph	17h 18m	-22° 16'	-12.2	32'	96%	156° E	FM	
4th	Sgr	18h 18m	-23° 50'	-12.5	32'	99%	170° E	FM	
5th	Sgr	19h 17m	-23° 56'	-12.6	31'	100%	176° W	FM	Jupiter
6th	Cap	20h 15m	-22° 39'	-12.3	31'	98%	162° W	FM	Jupiter, Saturn
7th	Cap	21h 10m	-20° 10'	-12.0	31'	94%	150° W	-G	
8th	Aqr	22h 2m	-16° 46'	-11.7	30'	88%	138° W	-G	
9th	Aqr	22h 50m	-12° 41'	-11.4	30'	81%	127° W	-G	Neptune
10th	Aqr	23h 36m	-8° 11'	-11.0	30'	73%	116° W	-G	Neptune

Mercury and Venus

 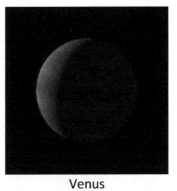

Mercury

5th

Venus
5th

Mercury

Date	Con.	R.A.	Dec.	Mag.	Diam.	Ill.	Elon.	Vis.	Rat.	Close To
1st	Gem	6h 40m	18° 36'	5.0	12"	1%	1° W	NV	N/A	
3rd	Gem	6h 35m	18° 29'	4.6	12"	1%	4° W	NV	N/A	
5th	Gem	6h 30m	18° 26'	4.1	12"	3%	7° W	NV	N/A	
7th	Gem	6h 27m	18° 29'	3.5	11"	5%	10° W	NV	N/A	
9th	Gem	6h 24m	18° 37'	2.9	11"	7%	13° W	NV	N/A	

Venus

Date	Con.	R.A.	Dec.	Mag.	Diam.	Ill.	Elon.	Vis.	Rat.	Close To
1st	Tau	4h 19m	17° 15'	-4.5	43"	19%	36° W	AM	****	Pleiades, Hyades, Aldebaran
3rd	Tau	4h 21m	17° 13'	-4.5	41"	21%	38° W	AM	****	Pleiades, Hyades, Aldebaran
5th	Tau	4h 24m	17° 13'	-4.5	40"	23%	39° W	AM	****	Pleiades, Hyades, Aldebaran
7th	Tau	4h 27m	17° 16'	-4.5	39"	25%	40° W	AM	****	Hyades, Aldebaran
9th	Tau	4h 31m	17° 21'	-4.5	38"	26%	41° W	AM	****	Hyades, Aldebaran

Mars and the Outer Planets

Mars
5th

Jupiter
5th

Saturn
5th

Mars

Date	Con.	R.A.	Dec.	Mag.	Diam.	Ill.	Elon.	Vis.	Rat.	Close To
1st	Psc	0h 12m	-2° 4'	-0.5	11"	84%	98° W	AM	***	
5th	Psc	0h 21m	-1° 13'	-0.6	12"	85%	100° W	AM	***	
10th	Cet	0h 32m	0° 12'	-0.7	12"	85%	102° W	AM	***	

The Outer Planets

Planet	Date	Con.	R.A.	Dec.	Mag.	Diam.	Elon.	Vis.	Rat.	Close To
Jupiter	5th	Sgr	19h 41m	-21° 42'	-2.7	47"	170° W	AM	*****	Moon, Saturn
Saturn	5th	Sgr	20h 7m	-20° 24'	0.2	18"	163° W	AM	****	
Uranus	5th	Ari	2h 31m	14° 23'	5.8	3"	67° W	AM	**	
Neptune	5th	Aqr	23h 28m	-4° 36'	7.9	2"	113° W	AM	****	

Highlights

Date	Time (UT)	Event
1st	02:47	Mercury is at inferior conjunction with the Sun. (Not visible.)
2nd	15:04	The waxing gibbous Moon is north of the bright star Antares. (Scorpius, evening sky.)
4th	19:56	Asteroid Vesta is in conjunction with the Sun. (Not visible.)
5th	04:29	Penumbral lunar eclipse. Visible from Africa, Antarctica, the Atlantic, Central America, western Europe, North America, the eastern Pacific and South America.
	04:45	Full Moon. (Visible all night.)
	20:32	The full Moon is south of Jupiter. (Visible all night.)
	21:20	The full Moon is south of dwarf planet Pluto. (Visible all night.)
6th	09:37	The just-past full Moon is south of Saturn. (Morning sky.)
10th	08:48	The waning gibbous Moon is south of Neptune. (Morning sky.)

Planet Locations – July 5th

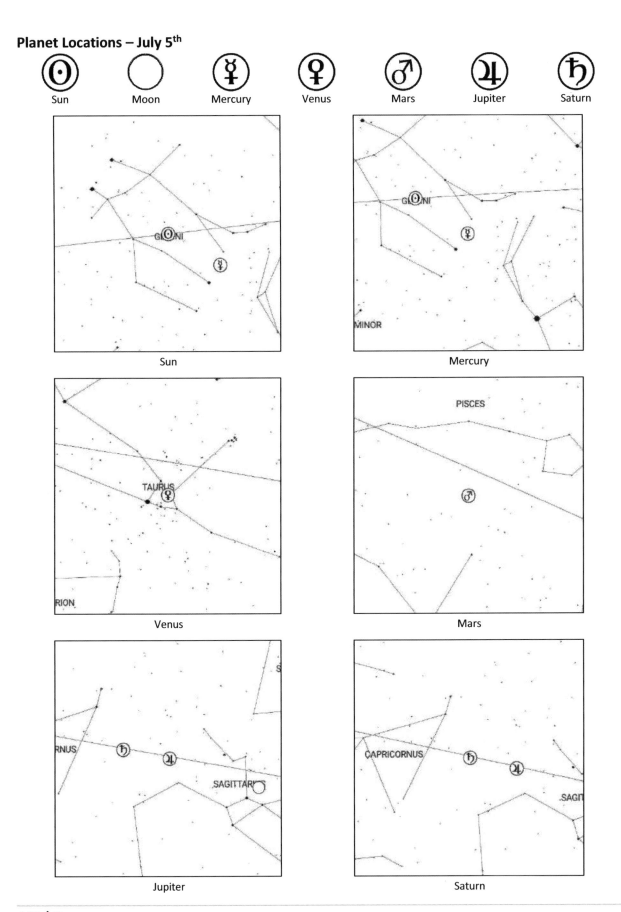

Sun   Moon   Mercury   Venus   Mars   Jupiter   Saturn

Sun

Mercury

Venus

Mars

Jupiter

Saturn

July 2ⁿᵈ – Evening Sky

There's a full Moon and penumbral lunar eclipse in two days time, but tonight the Moon is passing Antares, in Scorpius (Sco), the Scorpion.

A red giant star, Antares is so large that if it were placed at the center of our solar system it would extend beyond the orbit of Mars!

This image depicts the sky looking south at about two hours after sunset.

July 4ᵗʰ – Evening Sky

It's Independence Day in the United States, which means we can expect the sky to be lit by fireworks. Not normally a great night for astronomy, but tonight might be an exception.

The Moon turns full and there's a penumbral lunar eclipse, which reaches its maximum at 12:29 a.m. Eastern Time on the 5ᵗʰ. That's 9:29 p.m. Pacific Time on the evening of the 4ᵗʰ. Does the Moon appear a little dimmer tonight?

This image depicts the sky looking south-east at 11:29 p.m. Central Time.

July 5ᵗʰ – Evening Sky

The day after the lunar eclipse and the Moon can be found nestled with Jupiter and Saturn.

The trio won't rise until quite late, but if you wait until the morning they won't appear so close together.

This image depicts the sky looking south-east at 11:00 p.m.

July 11th to 20th, 2020

The Moon

11th

13th

15th

17th

19th

Date	Con	R.A.	Dec	Mag	Diam	Ill.	Elon.	Phase	Close To
11th	Psc	0h 20m	-3° 26'	-10.7	30'	64%	106° W	LQ	Mars
12th	Cet	1h 4m	1° 24'	-10.3	30'	55%	96° W	LQ	Mars
13th	Psc	1h 47m	6° 11'	-9.9	30'	45%	86° W	LQ	Uranus
14th	Ari	2h 32m	10° 45'	-9.4	30'	36%	76° W	-Cr	Uranus
15th	Ari	3h 18m	14° 57'	-8.9	30'	27%	66° W	-Cr	Uranus, Pleiades
16th	Tau	4h 7m	18° 37'	-8.3	30'	19%	54° W	-Cr	Venus, Pleiades, Hyades
17th	Tau	4h 59m	21° 31'	-7.5	31'	11%	42° W	NM	Venus, Hyades, Aldebaran
18th	Tau	5h 54m	23° 24'	-6.6	31'	6%	30° W	NM	Mercury
19th	Gem	6h 51m	24° 4'	-5.5	31'	2%	16° W	NM	Mercury
20th	Gem	7h 50m	23° 21'	-4.3	32'	0%	3° W	NM	

Mercury and Venus

Mercury
15th

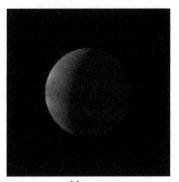
Venus
15th

Mercury

Date	Con.	R.A.	Dec.	Mag.	Diam.	Ill.	Elon.	Vis.	Rat.	Close To
11th	Gem	6h 23m	18° 50'	2.4	10"	11%	15° W	AM	**	
13th	Gem	6h 23m	19° 6'	1.9	10"	14%	17° W	AM	***	
15th	Gem	6h 25m	19° 25'	1.5	9"	19%	19° W	AM	***	
17th	Gem	6h 28m	19° 47'	1.1	9"	24%	20° W	AM	***	
19th	Gem	6h 33m	20° 9'	0.8	8"	29%	21° W	AM	***	Moon

Venus

Date	Con.	R.A.	Dec.	Mag.	Diam.	Ill.	Elon.	Vis.	Rat.	Close To
11th	Tau	4h 35m	17° 28'	-4.5	36"	28%	42° W	AM	****	Hyades, Aldebaran
13th	Tau	4h 39m	17° 36'	-4.5	35"	30%	43° W	AM	****	Hyades, Aldebaran
15th	Tau	4h 44m	17° 45'	-4.5	34"	31%	44° W	AM	****	Hyades, Aldebaran
17th	Tau	4h 49m	17° 55'	-4.5	33"	33%	45° W	AM	****	Moon, Hyades, Aldebaran
19th	Tau	4h 55m	18° 6'	-4.4	32"	34%	46° W	AM	****	Hyades, Aldebaran

Mars and the Outer Planets

Mars
15th

Jupiter
15th

Saturn
15th

Mars

Date	Con.	R.A.	Dec.	Mag.	Diam.	Ill.	Elon.	Vis.	Rat.	Close To
11th	Cet	0h 34m	0° 0'	-0.7	12"	85%	103° W	AM	***	Moon
15th	Cet	0h 42m	0° 46'	-0.8	13"	85%	105° W	AM	***	
20th	Cet	0h 52m	1° 42'	-0.9	13"	85%	107° W	AM	***	

The Outer Planets

Planet	Date	Con.	R.A.	Dec.	Mag.	Diam.	Elon.	Vis.	Rat.	Close To
Jupiter	15th	Sgr	19h 36m	-21° 56'	-2.8	48"	179° E	AN	*****	Saturn
Saturn	15th	Sgr	20h 4m	-20° 34'	0.1	18"	174° W	AN	*****	
Uranus	15th	Ari	2h 32m	14° 29'	5.8	4"	77° W	AM	**	Moon
Neptune	15th	Aqr	23h 28m	-4° 38'	7.9	2"	123° W	AM	****	

Highlights

Date	Time (UT)	Event
11th	17:51	Venus is 1.0° north of the bright star Aldebaran. (Taurus, morning sky.)
	19:00	The nearly last quarter Moon is south of Mars. (Morning sky.)
12th	06:48	Mercury is stationary prior to resuming prograde motion. (Morning sky.)
	23:30	Last Quarter Moon. (Morning sky.)
14th	09:06	Jupiter is at opposition. (Visible all night.)
	13:13	The waning crescent Moon is south of Uranus. (Morning sky.)
16th	00:49	The waning crescent Moon is south of the Pleiades star cluster. (Taurus, morning sky.)
17th	00:18	The waning crescent Moon is north of the bright star Aldebaran. (Taurus, morning sky.)
	06:35	The waning crescent Moon is north of Venus. (Morning sky.)
	N/A	Good opportunity to see Earthshine on the waning crescent Moon. (Morning sky.)
19th	02:48	The nearly new Moon is north of Mercury. (Morning sky.)
20th	17:33	New Moon. (Not visible.)
	23:33	Saturn is at opposition. (Visible all night.)

Planet Locations – July 15th

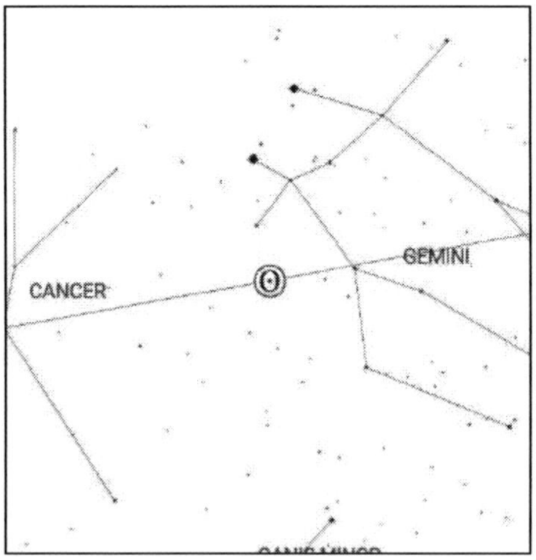

☉	☿	♀	♂	♃	♄
Sun	Mercury	Venus	Mars	Jupiter	Saturn

Sun

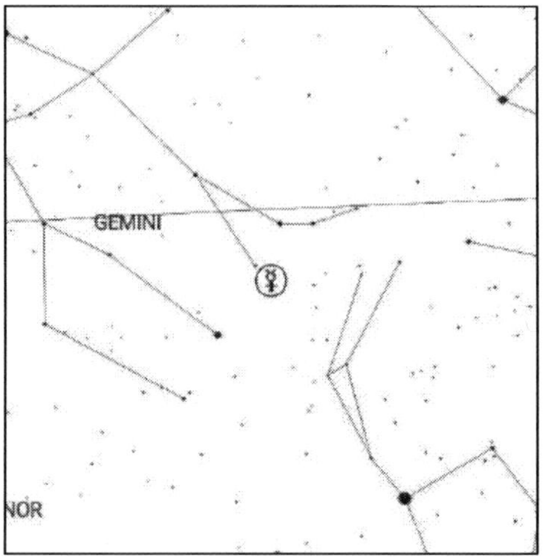

Mercury

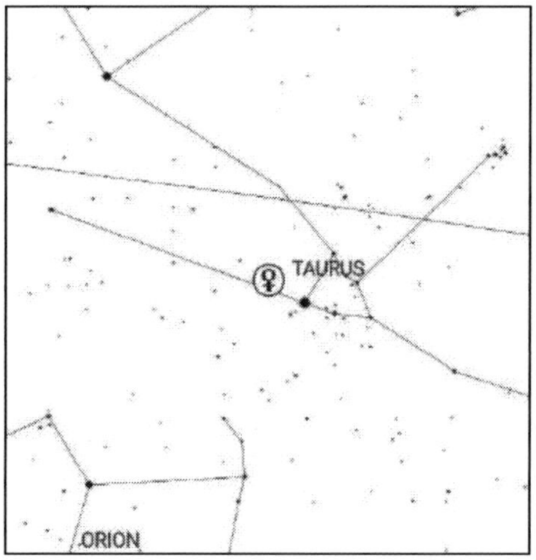

Venus

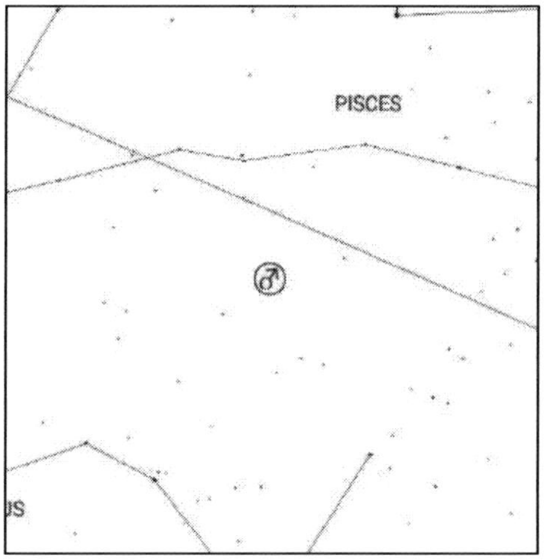

Mars

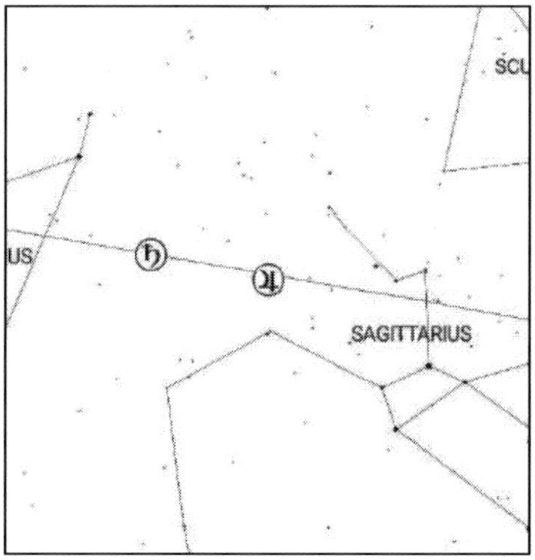

Jupiter

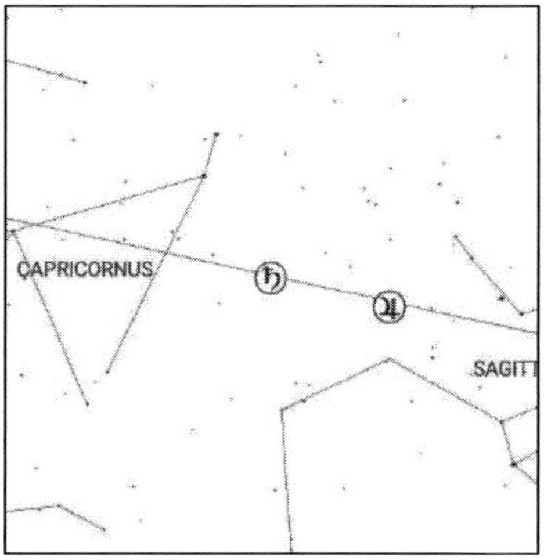

Saturn

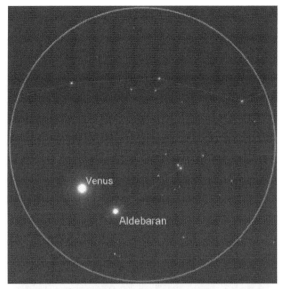

July 12th – Morning Sky

Venus has been dancing with the stars of Taurus (Tau), the Bull, for much of this year. Three months ago it appeared among the Pleiades star cluster in the evening sky.

It's since slipped into the pre-dawn sky and can now be found skirting by the V-shaped Hyades. This cluster forms the head of the bull, with orange Aldebaran marking his eye.

This image depicts the view through 10x50 binoculars looking east at about 75 minutes before sunrise.

July 14th – Morning Sky

Jupiter reaches opposition today, meaning that it's at its best visibility for the entire year. It rises at sunset, sets at sunrise and is therefore visible all night.

This July is prime planetary observing time for amateur astronomers. Jupiter reaches opposition today, but Saturn will also reach opposition in just another 6 days. Both planets are at their brightest and appear largest through a telescope.

This image depicts the sky looking south at 1:00 a.m.

July 17th – Morning Sky

If you've yet to see Mercury this year you may have another chance over the next week or so. The waning crescent Moon hovers close to Venus this morning and appears to point toward Mercury, now low over the eastern horizon.

Keep an eye on the Moon over the next few mornings and watch as it passes the tiny planet.

This image depicts the sky looking east about 30 minutes before sunrise.

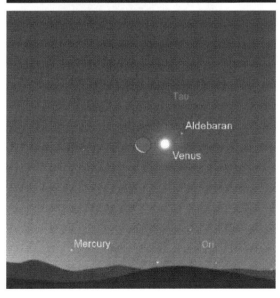

July 21ˢᵗ to 31ˢᵗ, 2020

The Moon

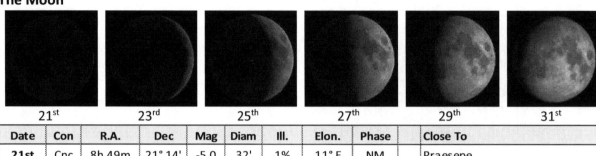

| 21ˢᵗ | 23ʳᵈ | 25ᵗʰ | 27ᵗʰ | 29ᵗʰ | 31ˢᵗ |

Date	Con	R.A.	Dec	Mag	Diam	Ill.	Elon.	Phase	Close To
21st	Cnc	8h 49m	21° 14'	-5.0	32'	1%	11° E	NM	Praesepe
22nd	Leo	9h 47m	17° 48'	-6.2	32'	4%	25° E	NM	Regulus
23rd	Leo	10h 43m	13° 18'	-7.3	32'	10%	38° E	NM	Regulus
24th	Leo	11h 37m	8° 2'	-8.2	32'	18%	50° E	+Cr	
25th	Vir	12h 30m	2° 19'	-8.9	32'	27%	62° E	+Cr	
26th	Vir	13h 22m	-3° 30'	-9.6	32'	38%	74° E	FQ	Spica
27th	Vir	14h 15m	-9° 6'	-10.1	32'	50%	87° E	FQ	
28th	Lib	15h 8m	-14° 11'	-10.6	32'	61%	99° E	FQ	
29th	Sco	16h 4m	-18° 27'	-11.0	32'	72%	112° E	+G	Antares
30th	Oph	17h 1m	-21° 39'	-11.4	32'	81%	125° E	+G	Antares
31st	Sgr	18h 0m	-23° 34'	-11.7	32'	89%	139° E	+G	

Mercury and Venus

Mercury
25ᵗʰ

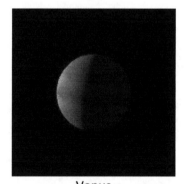

Venus
25ᵗʰ

Mercury

Date	Con.	R.A.	Dec.	Mag.	Diam.	Ill.	Elon.	Vis.	Rat.	Close To
21st	Gem	6h 39m	20° 32'	0.5	8"	34%	21° W	AM	***	
23rd	Gem	6h 47m	20° 52'	0.2	8"	41%	21° W	AM	***	
25th	Gem	6h 56m	21° 10'	-0.1	7"	47%	21° W	AM	***	
27th	Gem	7h 7m	21° 22'	-0.3	7"	54%	20° W	AM	***	
29th	Gem	7h 20m	21° 29'	-0.6	7"	61%	19° W	AM	***	
31st	Gem	7h 33m	21° 28'	-0.8	6"	68%	18° W	AM	**	

Venus

Date	Con.	R.A.	Dec.	Mag.	Diam.	Ill.	Elon.	Vis.	Rat.	Close To
21st	Tau	5h 0m	18° 18'	-4.4	31"	36%	46° W	AM	****	Hyades, Aldebaran
23rd	Tau	5h 6m	18° 29'	-4.4	30"	37%	47° W	AM	****	Hyades, Aldebaran
25th	Tau	5h 13m	18° 41'	-4.4	30"	39%	47° W	AM	****	Aldebaran
27th	Tau	5h 19m	18° 52'	-4.4	29"	40%	47° W	AM	****	
29th	Tau	5h 26m	19° 4'	-4.4	28"	41%	48° W	AM	****	
31st	Tau	5h 33m	19° 15'	-4.4	27"	43%	48° W	AM	****	

Mars and the Outer Planets

Mars
25th

Jupiter
25th

Saturn
25th

Mars

Date	Con.	R.A.	Dec.	Mag.	Diam.	Ill.	Elon.	Vis.	Rat.	Close To
21st	Cet	0h 54m	1° 52'	-0.9	13"	85%	108° W	AM	***	
25th	Cet	1h 1m	2° 34'	-1.0	14"	86%	110° W	AM	***	
31st	Psc	1h 12m	3° 31'	-1.1	15"	86%	113° W	AM	***	

The Outer Planets

Planet	Date	Con.	R.A.	Dec.	Mag.	Diam.	Elon.	Vis.	Rat.	Close To
Jupiter	25th	Sgr	19h 30m	-22° 9'	-2.7	47"	167° E	PM	*****	Saturn
Saturn	25th	Sgr	20h 1m	-20° 44'	0.1	18"	175° E	AN	*****	
Uranus	25th	Ari	2h 32m	14° 32'	5.8	4"	87° W	AM	**	
Neptune	25th	Aqr	23h 27m	-4° 42'	7.8	2"	133° W	AM	****	

Highlights

Date	Time (UT)	Event
22nd	15:04	Mercury is at greatest western elongation from the Sun. (Morning sky.)
	21:55	The waxing crescent Moon is north of the bright star Regulus. (Leo, evening sky.)
23rd	N/A	Good opportunity to see Earthshine on the waxing crescent Moon. (Evening sky.)
26th	11:56	The nearly first quarter Moon is north of the bright star Spica. (Virgo, evening sky.)
27th	12:33	First Quarter Moon. (Evening sky.)
28th	N/A	The Piscis Austrinid meteor shower is at its maximum. (ZHR: 5)
29th	23:42	The waxing gibbous Moon is north of the bright star Antares. (Scorpius, evening sky.)
30th	N/A	The Delta Aquariid meteor shower is at its maximum. (ZHR: 16)
	N/A	The Alpha Capricornid meteor shower is at its maximum. (ZHR: 5)

Planet Locations – July 25th

| Sun | Mercury | Venus | Mars | Jupiter | Saturn |

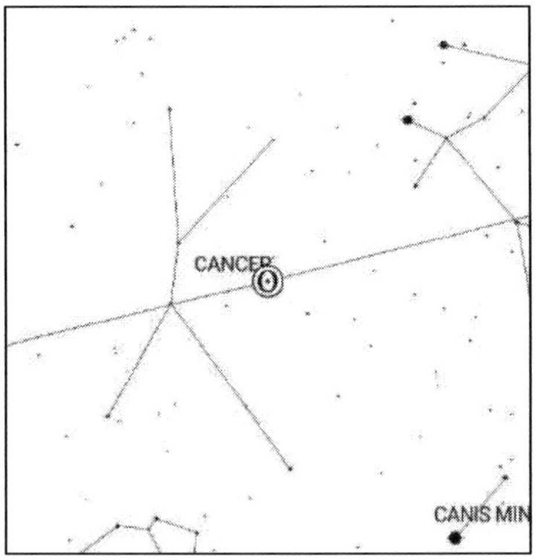

Sun

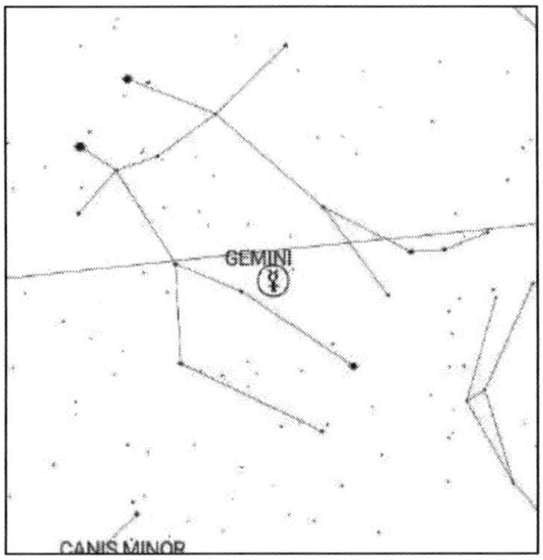

Mercury

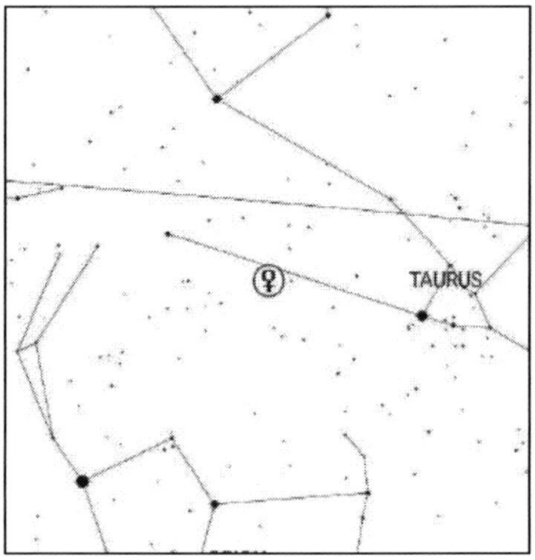

Venus

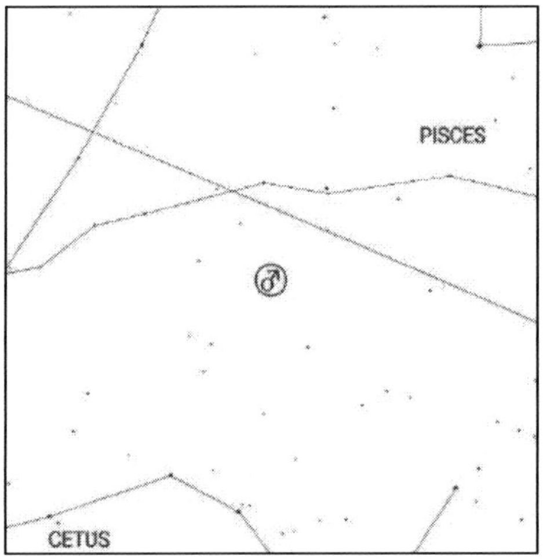

Mars

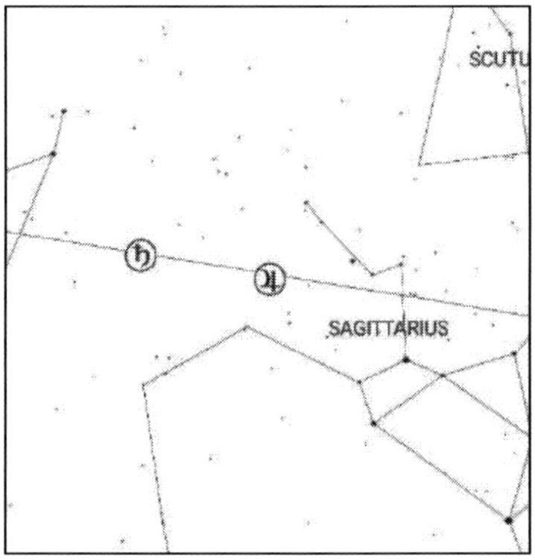

Jupiter

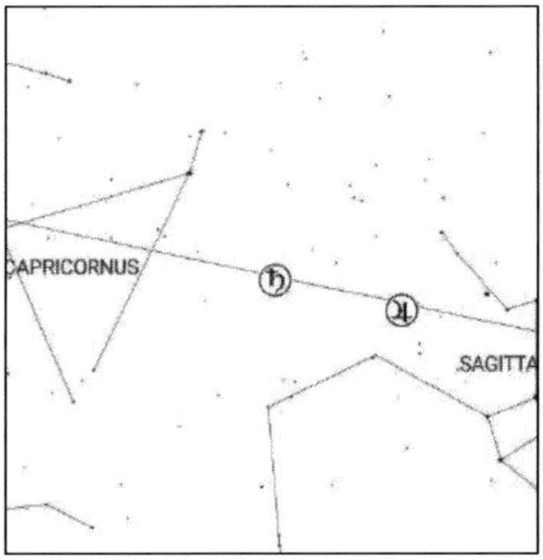

Saturn

July 22nd – Morning Sky

Just days after the waning crescent Moon passed the planet, Mercury is now at its best visibility in the morning sky.

If you have a clear, unobstructed view of the eastern horizon, you can catch it in the twilight before the dawn. Venus should be easily seen, but if you can also see the stars of Orion (Ori) they can help to point the way.

This image depicts the sky looking east at about 30 minutes before sunrise.

July 26th – Evening Sky

The Moon reaches first quarter tomorrow but you'll find it among the stars of spring tonight.

Look for it close to Spica, the brightest star in the constellation of Virgo (Vir), the Virgin. Faint Libra (Lib) the Scales, lies to the east. Can you see any of its stars against the darkening sky?

This image depicts the sky looking south-west at about an hour after sunset.

July 29th – Evening Sky

Having left the spring stars behind, the Moon now skirts along the southerly constellations of summer. Tonight you'll find the waxing gibbous Moon to the upper left of Antares.

The remaining stars of Scorpius (Sco) the Scorpion appear below, while Sagittarius (Sgr) the Archer, is slowly emerging into the twilight toward the south-east.

This image depicts the sky looking south at about an hour after sunset.

August 1st to 10th, 2020

The Moon

| 1st | 3rd | 5th | 7th | 9th |

Date	Con	R.A.	Dec	Mag	Diam	Ill.	Elon.	Phase	Close To
1st	Sgr	18h 58m	-24° 5'	-12.0	31'	95%	153° E	+G	Jupiter
2nd	Sgr	19h 56m	-23° 13'	-12.4	31'	98%	166° E	FM	Jupiter, Saturn
3rd	Cap	20h 52m	-21° 6'	-12.6	31'	100%	179° E	FM	
4th	Cap	21h 44m	-17° 58'	-12.5	30'	99%	169° W	FM	
5th	Aqr	22h 34m	-14° 4'	-12.2	30'	96%	157° W	FM	
6th	Aqr	23h 21m	-9° 39'	-11.9	30'	92%	147° W	-G	Neptune
7th	Psc	0h 6m	-4° 55'	-11.6	30'	86%	136° W	-G	Neptune
8th	Cet	0h 49m	0° 3'	-11.3	30'	79%	126° W	-G	Mars
9th	Psc	1h 33m	4° 46'	-10.9	30'	70%	116° W	-G	Mars
10th	Cet	2h 17m	9° 25'	-10.6	30'	61%	106° W	LQ	Uranus

Mercury and Venus

Mercury
5th

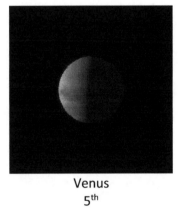

Venus
5th

Mercury

Date	Con.	R.A.	Dec.	Mag.	Diam.	Ill.	Elon.	Vis.	Rat.	Close To
1st	Gem	7h 41m	21° 24'	-0.9	6"	71%	17° W	AM	**	
3rd	Gem	7h 56m	21° 10'	-1.0	6"	78%	15° W	NV	N/A	
5th	Cnc	8h 12m	20° 45'	-1.2	6"	84%	13° W	NV	N/A	Praesepe
7th	Cnc	8h 28m	20° 9'	-1.3	5"	89%	11° W	NV	N/A	Praesepe
9th	Cnc	8h 45m	19° 23'	-1.5	5"	93%	8° W	NV	N/A	Praesepe

Venus

Date	Con.	R.A.	Dec.	Mag.	Diam.	Ill.	Elon.	Vis.	Rat.	Close To
1st	Tau	5h 37m	19° 20'	-4.4	27"	43%	48° W	AM	****	
3rd	Tau	5h 44m	19° 29'	-4.4	26"	45%	48° W	AM	****	
5th	Ori	5h 51m	19° 38'	-4.4	26"	46%	48° W	AM	****	
7th	Ori	5h 59m	19° 46'	-4.3	25"	47%	48° W	AM	***	
9th	Ori	6h 7m	19° 53'	-4.3	25"	48%	48° W	AM	***	

Mars and the Outer Planets

Mars
5th

Jupiter
5th

Saturn
5th

Mars

Date	Con.	R.A.	Dec.	Mag.	Diam.	Ill.	Elon.	Vis.	Rat.	Close To
1st	Psc	1h 13m	3° 40'	-1.1	15"	86%	114° W	AM	***	
5th	Psc	1h 20m	4° 14'	-1.2	15"	87%	116° W	AM	***	
10th	Psc	1h 27m	4° 52'	-1.3	16"	88%	119° W	AM	****	

The Outer Planets

Planet	Date	Con.	R.A.	Dec.	Mag.	Diam.	Elon.	Vis.	Rat.	Close To
Jupiter	5th	Sgr	19h 25m	-22° 22'	-2.7	47"	155° E	PM	*****	Saturn
Saturn	5th	Sgr	19h 58m	-20° 54'	0.2	18"	164° E	PM	****	
Uranus	5th	Ari	2h 33m	14° 35'	5.8	4"	98° W	AM	**	
Neptune	5th	Aqr	23h 26m	-4° 47'	7.8	2"	144° W	AM	****	

Highlights

Date	Time (UT)	Event
1st	23:55	The waxing gibbous Moon is south of Jupiter. (Evening sky.)
2nd	06:13	The nearly full Moon is south of dwarf planet Pluto. (Evening sky.)
	12:41	The nearly full Moon is south of Saturn. (Evening sky.)
3rd	15:59	Full Moon (Visible all night.)
6th	14:44	The waning gibbous Moon is south of Neptune. (Morning sky.)
9th	09:26	The waning gibbous Moon is south of Mars. (Morning sky.)
10th	19:54	The nearly last quarter Moon is south of Uranus. (Morning sky.)

Planet Locations – August 5th

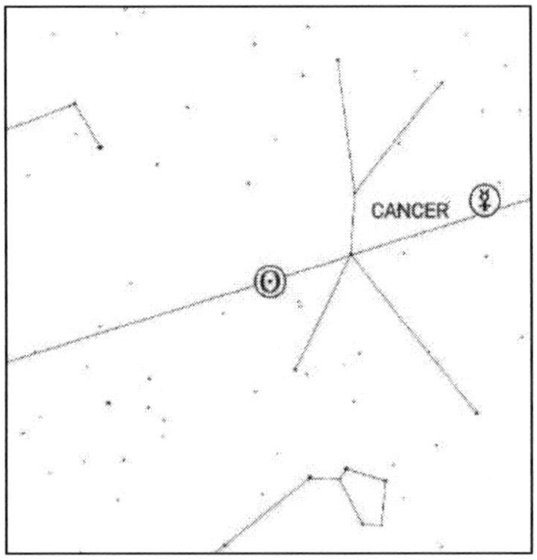

Sun 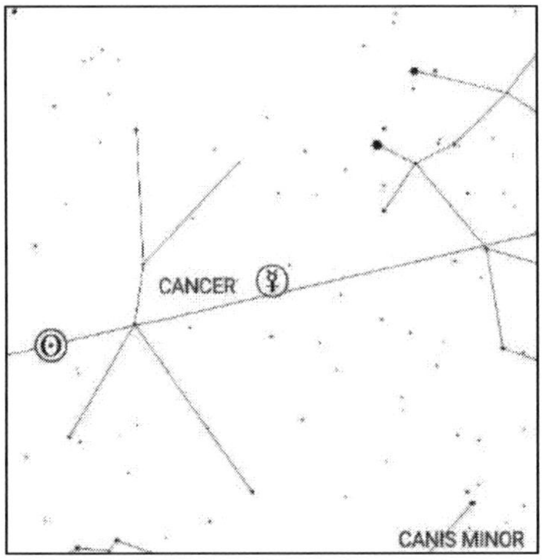   Mercury 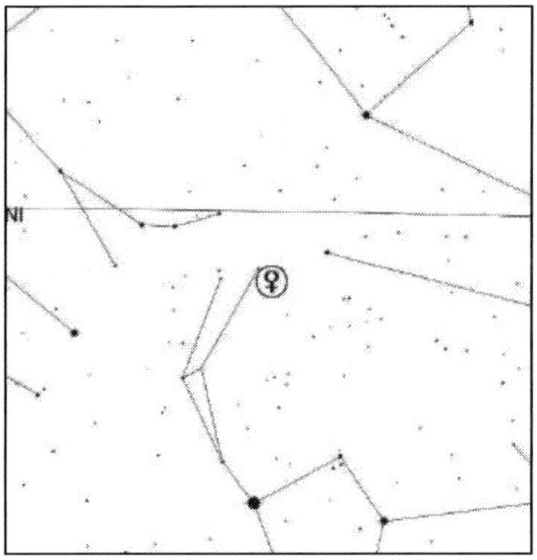   Venus 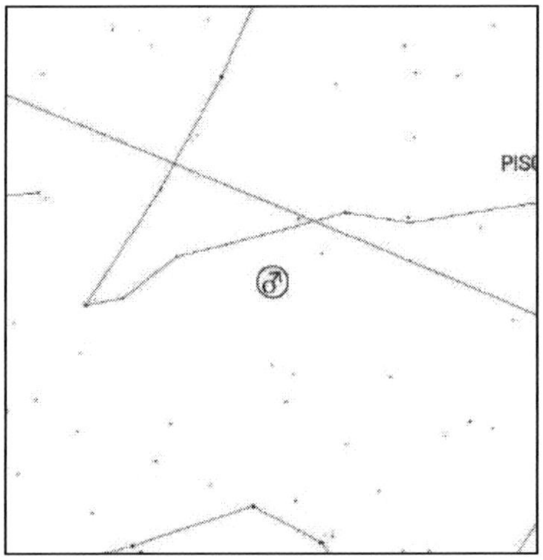   Mars 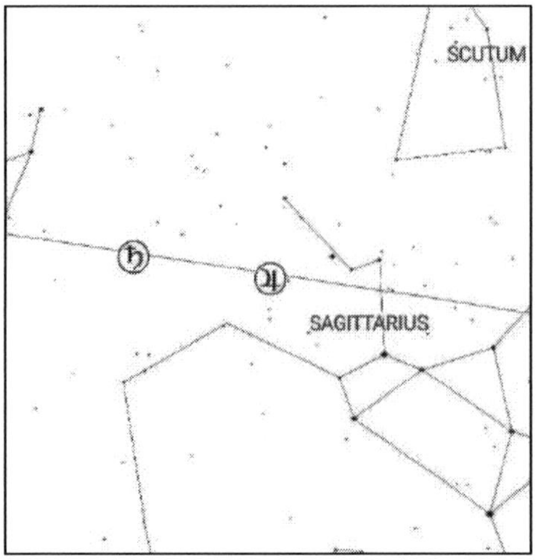   Jupiter   Saturn

Sun

Mercury

Venus

Mars

Jupiter

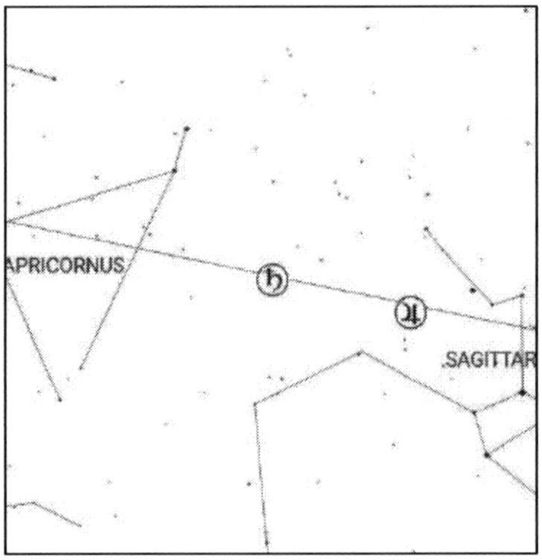

Saturn

August 1st – Evening Sky

The Moon is nearly full and its light is drowning out all but the brightest stars. Fortunately, the planets Jupiter and Saturn can still be clearly seen.

Jupiter hangs to the upper right of the Moon while Saturn appears to the left. The Moon will skip past the ringed planet over the next 24 hours and will be very close to Mars in about 8 days time.

This image depicts the sky looking south-east at about 90 minutes after sunset.

August 5th – Evening Sky

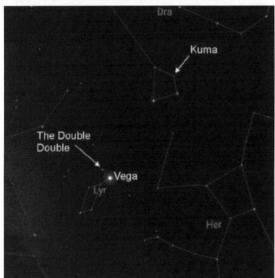

If you own a pair of binoculars, there's a couple of easy multiple stars you can enjoy tonight. Both appear as twin white stars of equal brightness.

The famous Double Double lies within the same field of view as Vega, the brightest star in Lyra (Lyr), the Lyre. There are actually four stars here (hence the name) but you can only see two with binoculars. Kuma is a little trickier to find but is one of the stars that forms the head of Draco (Dra), the Dragon.

This image depicts the sky looking overhead at about two hours after sunset.

August 9th – Morning Sky

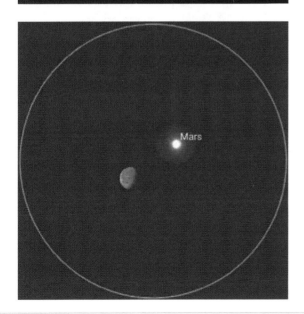

There's a nice opportunity to catch the waning gibbous Moon and Mars close together in the pre-dawn sky this morning.

There's about a degree and a half between the pair, so you'll easily be able to fit them within the same binocular field of view.

This image depicts the view through 10x50 binoculars looking south at about an hour before sunrise.

August 11th to 20th, 2020

The Moon

| 11th | 13th | 15th | 17th | 19th |

Date	Con	R.A.	Dec	Mag	Diam	Ill.	Elon.	Phase	Close To
11th	Ari	3h 2m	13° 44'	-10.2	30'	52%	96° W	LQ	Uranus, Pleiades
12th	Tau	3h 49m	17° 33'	-9.8	30'	42%	85° W	LQ	Pleiades, Hyades, Aldebaran
13th	Tau	4h 39m	20° 42'	-9.3	30'	33%	74° W	-Cr	Hyades, Aldebaran
14th	Tau	5h 33m	22° 57'	-8.7	31'	24%	61° W	-Cr	
15th	Gem	6h 29m	24° 4'	-8.0	31'	16%	48° W	-Cr	Venus
16th	Gem	7h 27m	23° 52'	-7.1	32'	9%	34° W	NM	
17th	Cnc	8h 26m	22° 15'	-6.1	32'	4%	21° W	NM	Praesepe
18th	Leo	9h 25m	19° 13'	-4.9	32'	1%	7° W	NM	Mercury, Praesepe, Regulus
19th	Leo	10h 23m	14° 58'	-4.6	33'	0%	7° E	NM	Mercury, Regulus
20th	Leo	11h 19m	9° 45'	-5.9	33'	3%	20° E	NM	

Mercury and Venus

Mercury
15th

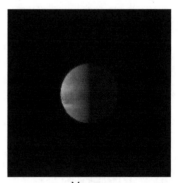

Venus
15th

Mercury

Date	Con.	R.A.	Dec.	Mag.	Diam.	Ill.	Elon.	Vis.	Rat.	Close To
11th	Cnc	9h 2m	18° 27'	-1.6	5"	96%	6° W	NV	N/A	Praesepe
13th	Cnc	9h 18m	17° 23'	-1.7	5"	98%	4° W	NV	N/A	Praesepe
15th	Leo	9h 35m	16° 11'	-1.9	5"	99%	2° W	NV	N/A	Regulus
17th	Leo	9h 51m	14° 53'	-1.9	5"	100%	0° E	NV	N/A	Regulus
19th	Leo	10h 6m	13° 30'	-1.7	5"	100%	2° E	NV	N/A	Moon, Regulus

Venus

Date	Con.	R.A.	Dec.	Mag.	Diam.	Ill.	Elon.	Vis.	Rat.	Close To
11th	Ori	6h 15m	19° 59'	-4.3	24"	49%	48° W	AM	***	
13th	Gem	6h 23m	20° 3'	-4.3	23"	50%	48° W	AM	***	
15th	Gem	6h 31m	20° 5'	-4.3	23"	51%	48° W	AM	***	Moon
17th	Gem	6h 39m	20° 7'	-4.3	22"	52%	47° W	AM	***	
19th	Gem	6h 48m	20° 6'	-4.3	22"	54%	47° W	AM	***	

Mars and the Outer Planets

Mars
15th

Jupiter
15th

Saturn
15th

Mars

Date	Con.	R.A.	Dec.	Mag.	Diam.	Ill.	Elon.	Vis.	Rat.	Close To
11th	Psc	1h 29m	4° 59'	-1.3	16"	88%	119° W	AM	****	
15th	Psc	1h 34m	5° 26'	-1.4	16"	88%	122° W	AM	****	
20th	Psc	1h 39m	5° 54'	-1.5	17"	89%	125° W	AM	****	

The Outer Planets

Planet	Date	Con.	R.A.	Dec.	Mag.	Diam.	Elon.	Vis.	Rat.	Close To
Jupiter	15th	Sgr	19h 21m	-22° 31'	-2.7	46"	145° E	PM	****	Saturn
Saturn	15th	Sgr	19h 55m	-21° 3'	0.2	18"	153° E	PM	****	
Uranus	15th	Ari	2h 33m	14° 35'	5.7	4"	107° W	AM	***	
Neptune	15th	Aqr	23h 25m	-4° 52'	7.8	2"	154° W	AM	*****	

Highlights

Date	Time (UT)	Event
11th	16:45	Last Quarter Moon. (Morning sky.)
12th	11:52	The just-past last quarter Moon is south of the Pleaides star cluster. (Taurus, morning sky.)
	23:49	Venus is at greatest western elongation from the Sun. (Morning sky.)
13th	11:11	The waning crescent Moon is north of the bright star Aldebaran. (Taurus, morning sky.)
	N/A	The Perseid meteor shower is at its maximum. (ZHR: 100)
15th	13:27	Uranus is stationary prior to beginning retrograde motion. (Morning sky.)
	14:18	The waning crescent Moon is north of Venus. (Morning sky.)
	N/A	Good opportunity to see Earthshine on the waning crescent Moon. (Morning sky.)
17th	14:54	Mercury is at superior conjunction with the Sun. (Not visible.)
18th	N/A	The Kappa Cygnid meteor shower is at its maximum. (ZHR: 3)
19th	02:42	New Moon. (Not visible.)

Planet Locations – August 15th

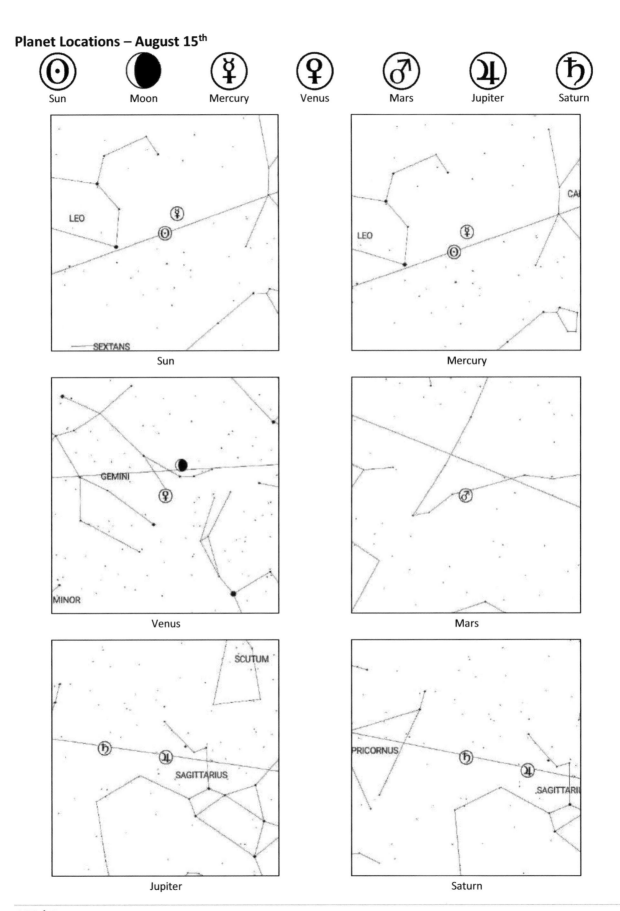

Planet Locations – August 15th

August 12th – Morning Sky

Venus reaches greatest western elongation from the Sun today. This means it appears at its furthest distance from the Sun in the sky and is best seen in the pre-dawn twilight sky.

You'll find it among the stars of winter with Gemini (Gem) to its lower left and Procyon, the brightest star in Canis Minor (CMi) the Little Dog, directly below it.

This image depicts the sky looking east at about 30 minutes before sunrise.

August 12th to 13th – Evening & Morning Sky

The Perseid meteor shower reaches its maximum in the early hours of this morning. You can, of course, see shooting stars late in the evening of the 12th, but you'll see more once the constellation of Perseus (Per) rises around 11 p.m.

Fortunately, the Moon is out of the way and its light won't interfere too much. It's currently close to the bright star Aldebaran and won't rise until around 2 a.m.

This image depicts the sky looking east at about 2:00 a.m.

August 15th – Morning Sky

Venus continues to shine brilliantly above the south-eastern horizon in the pre-dawn sky. This morning it's joined by the waning crescent Moon, which appears to the upper left of our nearest planetary neighbor.

This is also a good opportunity to see Earthshine illuminating the darkened lunar surface and should make for a very nice view.

This image depicts the sky looking east at about 30 minutes before sunrise.

August 21st to 31st, 2020

The Moon

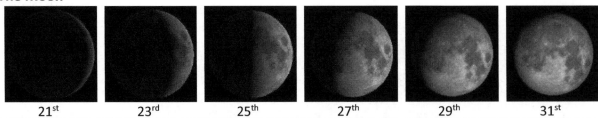

| | 21st | | 23rd | | 25th | | 27th | | 29th | | 31st |

Date	Con	R.A.	Dec	Mag	Diam	Ill.	Elon.	Phase	Close To
21st	Vir	12h 14m	3° 57'	-7.1	33'	8%	33° E	NM	
22nd	Vir	13h 7m	-2° 3'	-8.0	33'	16%	45° E	+Cr	Spica
23rd	Vir	14h 1m	-7° 54'	-8.8	33'	25%	58° E	+Cr	Spica
24th	Lib	14h 55m	-13° 15'	-9.5	32'	36%	70° E	+Cr	
25th	Lib	15h 51m	-17° 46'	-10.0	32'	47%	83° E	FQ	Antares
26th	Oph	16h 48m	-21° 13'	-10.5	32'	58%	96° E	FQ	Antares
27th	Sgr	17h 46m	-23° 24'	-10.9	31'	69%	110° E	+G	
28th	Sgr	18h 44m	-24° 13'	-11.2	31'	78%	124° E	+G	Jupiter
29th	Sgr	19h 41m	-23° 40'	-11.6	31'	86%	137° E	+G	Jupiter, Saturn
30th	Cap	20h 37m	-21° 51'	-11.9	31'	93%	150° E	+G	Saturn
31st	Cap	21h 29m	-18° 59'	-12.2	30'	97%	162° E	FM	

Mercury and Venus

Mercury
25th

Venus
25th

Mercury

Date	Con.	R.A.	Dec.	Mag.	Diam.	Ill.	Elon.	Vis.	Rat.	Close To
21st	Leo	10h 21m	12° 3'	-1.5	5"	99%	4° E	NV	N/A	Regulus
23rd	Leo	10h 35m	10° 34'	-1.3	5"	98%	6° E	NV	N/A	Regulus
25th	Leo	10h 49m	9° 2'	-1.1	5"	97%	8° E	NV	N/A	
27th	Leo	11h 3m	7° 30'	-0.9	5"	95%	9° E	NV	N/A	
29th	Leo	11h 16m	5° 57'	-0.8	5"	94%	11° E	NV	N/A	
31st	Leo	11h 28m	4° 24'	-0.7	5"	93%	12° E	NV	N/A	

Venus

Date	Con.	R.A.	Dec.	Mag.	Diam.	Ill.	Elon.	Vis.	Rat.	Close To
21st	Gem	6h 57m	20° 4'	-4.3	22"	55%	47° W	AM	***	
23rd	Gem	7h 5m	20° 0'	-4.2	21"	56%	46° W	AM	***	
25th	Gem	7h 14m	19° 54'	-4.2	21"	56%	46° W	AM	***	
27th	Gem	7h 23m	19° 47'	-4.2	20"	57%	46° W	AM	***	
29th	Gem	7h 32m	19° 37'	-4.2	20"	58%	45° W	AM	***	
31st	Gem	7h 41m	19° 26'	-4.2	20"	59%	45° W	AM	***	

Mars and the Outer Planets

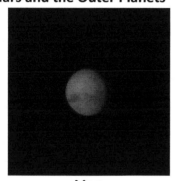

Mars
25th

Jupiter
25th

Saturn
25th

Mars

Date	Con.	R.A.	Dec.	Mag.	Diam.	Ill.	Elon.	Vis.	Rat.	Close To
21st	Psc	1h 40m	5° 59'	-1.5	17"	90%	126° W	AM	****	
25th	Psc	1h 44m	6° 17'	-1.6	18"	90%	129° W	AM	****	
31st	Psc	1h 48m	6° 37'	-1.8	19"	92%	133° W	AM	****	

The Outer Planets

Planet	Date	Con.	R.A.	Dec.	Mag.	Diam.	Elon.	Vis.	Rat.	Close To
Jupiter	25th	Sgr	19h 18m	-22° 38'	-2.6	45"	135° E	PM	****	Saturn
Saturn	25th	Sgr	19h 53m	-21° 10'	0.3	18"	144° E	PM	****	
Uranus	25th	Ari	2h 33m	14° 35'	5.7	4"	116° W	AM	***	
Neptune	25th	Aqr	23h 25m	-4° 59'	7.8	2"	163° W	AM	*****	

Highlights

Date	Time (UT)	Event
22nd	21:01	The waxing crescent Moon is north of the bright star Spica. (Virgo, evening sky.)
	N/A	Good opportunity to see Earthshine on the waxing crescent Moon. (Evening sky.)
25th	17:58	First Quarter Moon. (Evening sky.)
26th	04:33	The just-past first quarter Moon is north of Antares. (Scorpius, evening sky.)
29th	02:50	The waxing gibbous Moon is south of Jupiter. (Evening sky.)
	10:01	The waxing gibbous Moon is south of the dwarf planet Pluto. (Evening sky.)
	15:18	The waxing gibbous Moon is south of Saturn. (Evening sky.)

Planet Locations – August 25th

☉	☿	♀	♂	♃	♄
Sun	Mercury	Venus	Mars	Jupiter	Saturn

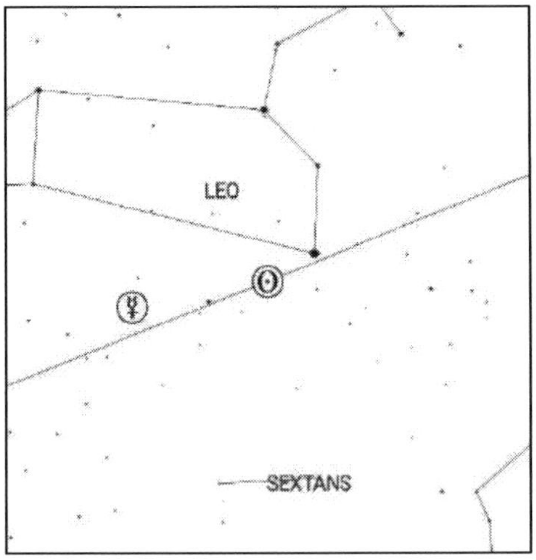

Sun

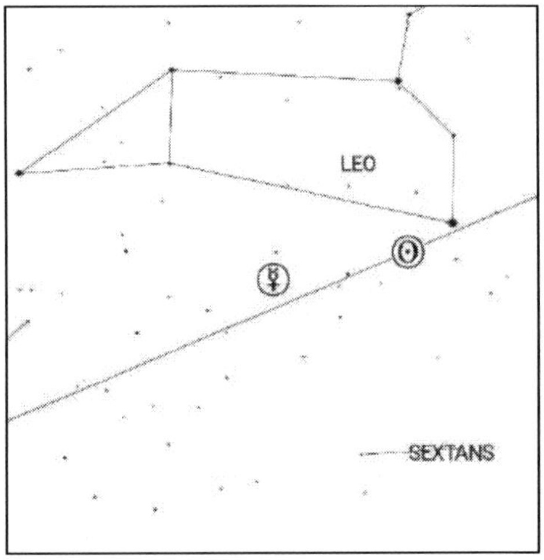

Mercury

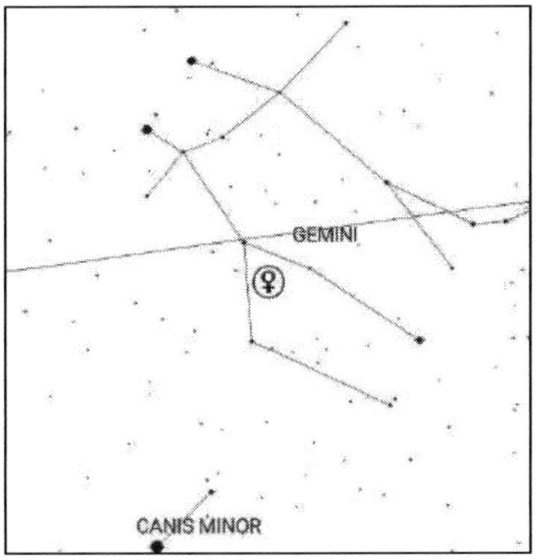

Venus

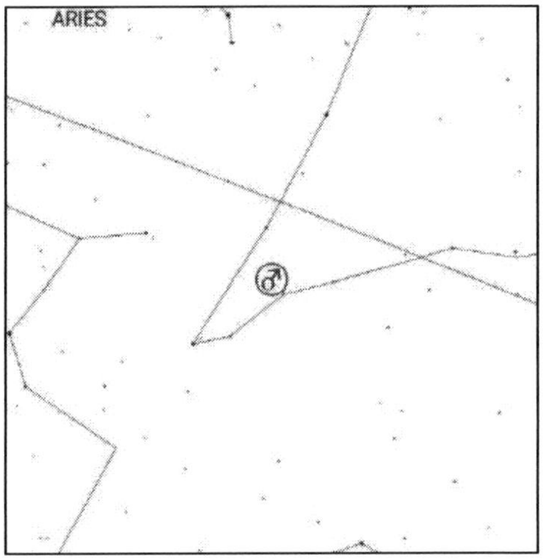

Mars

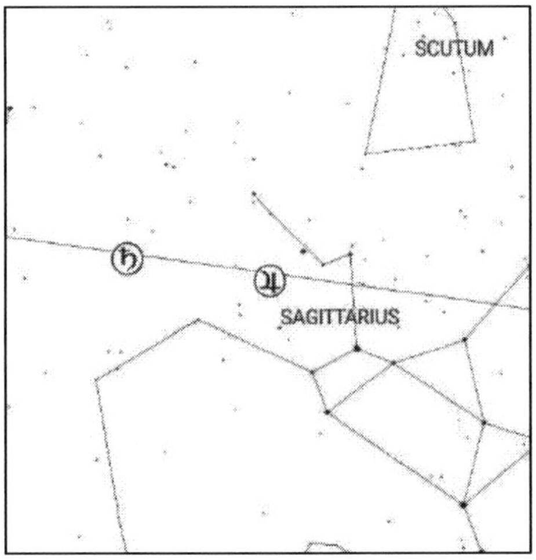

Jupiter

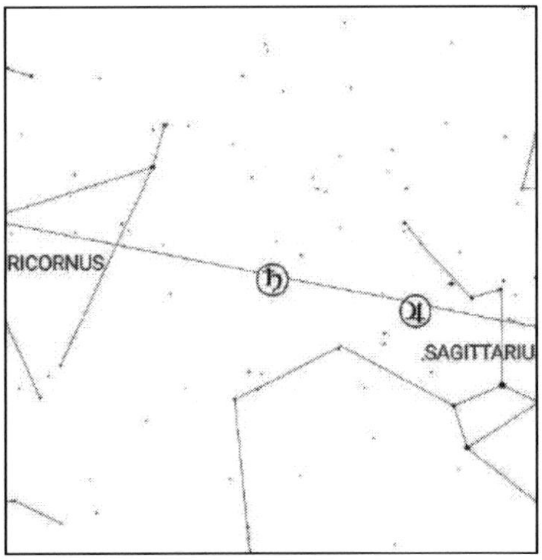

Saturn

August 22nd – Evening Sky

After turning new three days ago, the Moon is paying one last visit to the spring star Spica. Now in its waxing crescent phase, it can be seen to the upper right of the star in the evening twilight.

Earthshine can also be seen on the lunar surface, making the whole of the Moon visible.

This image depicts the sky looking south-west about 45 minutes after sunset.

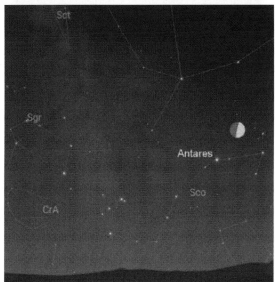

August 25th – Evening Sky

August also provides us with one last chance to easily see the Moon close to Antares, the beating red heart of Scorpius (Sco) the Scorpion.

The Moon has now reached first quarter and can be found just to the upper right of the bright star.

This image depicts the sky looking south at about 75 minutes after sunset.

August 28th – Evening Sky

The giant worlds of Jupiter and Saturn are drawing closer together and can be seen until the end of the year.

They're currently visible in the evening sky and are joined by the waxing gibbous Moon. It's below Jupiter tonight and will appear on the other side of Saturn by this time tomorrow.

This image depicts the sky looking south at about an hour and 45 minutes after sunset.

September 1st to 10th, 2020

The Moon

1st

3rd

5th

7th

9th

Date	Con	R.A.	Dec	Mag	Diam	Ill.	Elon.	Phase	Close To
1st	Aqr	22h 19m	-15° 16'	-12.5	30'	99%	174° E	FM	
2nd	Aqr	23h 7m	-10° 58'	-12.6	30'	100%	175° W	FM	Neptune
3rd	Aqr	23h 52m	-6° 16'	-12.3	30'	98%	165° W	FM	Neptune
4th	Cet	0h 36m	-1° 24'	-12.1	30'	95%	155° W	-G	
5th	Psc	1h 19m	3° 29'	-11.8	29'	90%	145° W	-G	Mars
6th	Psc	2h 3m	8° 14'	-11.5	29'	84%	135° W	-G	Mars, Uranus
7th	Ari	2h 48m	12° 40'	-11.2	30'	76%	124° W	-G	Uranus
8th	Tau	3h 34m	16° 39'	-10.8	30'	68%	114° W	-G	Pleiades
9th	Tau	4h 22m	19° 59'	-10.5	30'	59%	103° W	LQ	Pleiades, Hyades, Aldebaran
10th	Tau	5h 14m	22° 31'	-10.1	30'	49%	91° W	LQ	Hyades, Aldebaran

Mercury and Venus

Mercury
5th

Venus
5th

Mercury

Date	Con.	R.A.	Dec.	Mag.	Diam.	Ill.	Elon.	Vis.	Rat.	Close To
1st	Leo	11h 35m	3° 38'	-0.6	5"	92%	13° E	NV	N/A	
3rd	Vir	11h 47m	2° 6'	-0.5	5"	90%	14° E	NV	N/A	
5th	Vir	11h 58m	0° 35'	-0.4	5"	89%	15° E	NV	N/A	
7th	Vir	12h 10m	0° 55'	-0.3	5"	87%	16° E	PM	**	
9th	Vir	12h 21m	-2° 23'	-0.3	5"	85%	17° E	PM	**	

Venus

Date	Con.	R.A.	Dec.	Mag.	Diam.	Ill.	Elon.	Vis.	Rat.	Close To
1st	Gem	7h 45m	19° 19'	-4.2	19"	60%	45° W	AM	***	
3rd	Gem	7h 54m	19° 5'	-4.2	19"	61%	44° W	AM	***	
5th	Cnc	8h 3m	18° 48'	-4.2	19"	62%	44° W	AM	***	Praesepe
7th	Cnc	8h 12m	18° 30'	-4.2	18"	62%	43° W	AM	***	Praesepe
9th	Cnc	8h 21m	18° 9'	-4.2	18"	63%	43° W	AM	***	Praesepe

Mars and the Outer Planets

Mars
5th

Jupiter
5th

Saturn
5th

Mars

Date	Con.	R.A.	Dec.	Mag.	Diam.	Ill.	Elon.	Vis.	Rat.	Close To
1st	Psc	1h 48m	6° 39'	-1.8	19"	92%	134° W	AM	****	
5th	Psc	1h 50m	6° 47'	-1.9	20"	93%	137° W	AM	****	Moon
10th	Psc	1h 50m	6° 51'	-2.0	20"	94%	142° W	AM	****	

The Outer Planets

Planet	Date	Con.	R.A.	Dec.	Mag.	Diam.	Elon.	Vis.	Rat.	Close To
Jupiter	5th	Sgr	19h 16m	-22° 42'	-2.5	44"	124° E	PM	****	Saturn
Saturn	5th	Sgr	19h 51m	-21° 16'	0.3	18"	133° E	PM	****	
Uranus	5th	Ari	2h 32m	14° 32'	5.7	4"	126° W	AM	***	
Neptune	5th	Aqr	23h 23m	-5° 6'	7.8	2"	174° W	AN	*****	

Highlights

Date	Time (UT)	Event
1st	N/A	The Alpha Aurigid meteor shower is at its maximum. (ZHR: 6)
2nd	05:23	Full Moon. (Visible all night.)
	18:14	Dwarf planet Ceres is at opposition. (Visible all night.)
	19:34	The full Moon is south of Neptune. (Visible all night.)
6th	05:45	The waning gibbous Moon is south of Mars. (Morning sky.)
7th	04:07	The waning gibbous Moon is south of Uranus. (Morning sky.)
8th	17:52	The waning gibbous Moon is south of the Pleaides star cluster. (Taurus, morning sky.)
9th	17:56	The nearly last quarter Moon is north of Aldebaran. (Taurus, morning sky.)
	18:04	Mars is stationary prior to beginning retrograde motion. (Morning sky.)
10th	09:26	Last Quarter Moon. (Morning sky.)
	N/A	The Epsilon Perseid meteor shower is at its maximum. (ZHR: 5)

Planet Locations – September 5th

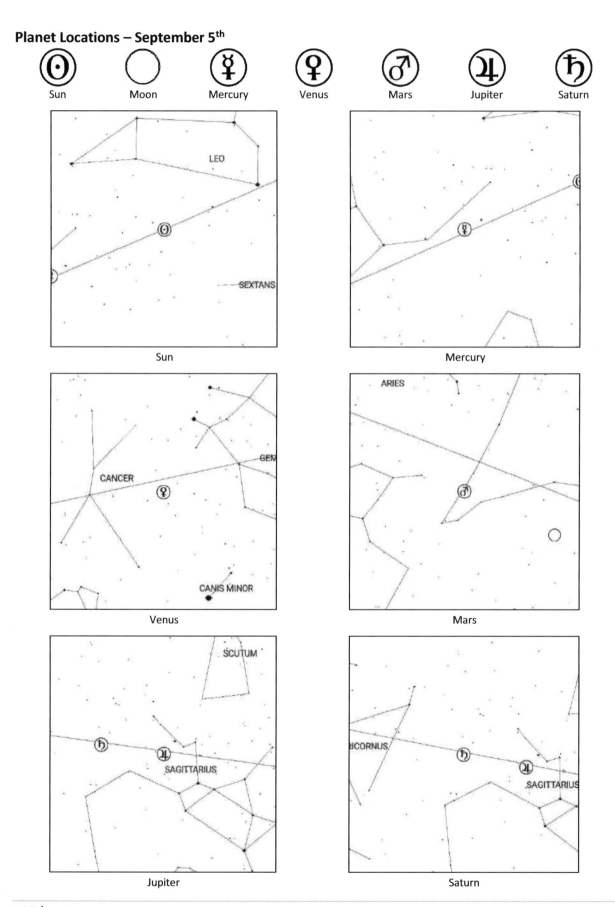

Sun	Moon	Mercury	Venus	Mars	Jupiter	Saturn

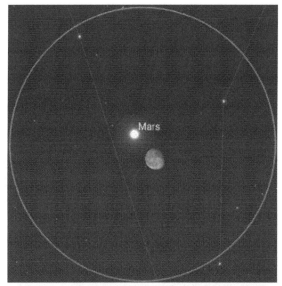

September 5th – Evening Sky

Despite turning full on the 2nd, the Moon is still visible in the evening sky and continues to rise in the hours before midnight. Tonight, once again, it appears very close to the planet Mars.

Now in its waning gibbous phase, it can be found to the lower right of the planet and both will easily fit within the same binocular field of view.

This image depicts the view through 10x50 binoculars looking east at about 10:30 p.m.

September 9th - Morning Sky

The Moon reaches last quarter in the early hours of tomorrow morning, but you'll find it close to Aldebaran in the pre-dawn skies today.

It is, of course, currently waning and, as a result, you may be able to see more stars nearby. Can you see the V-shaped Hyades that form the head of Taurus (Tau), the Bull? You'll find them directly below the Moon this morning.

This image depicts the sky looking south-east at about 75 minutes before sunrise.

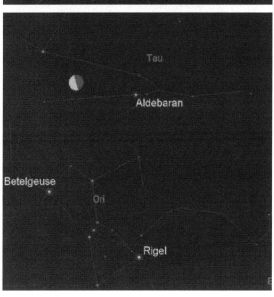

September 10th – Morning Sky

The Moon reached last quarter a few hours ago and has moved further east since yesterday morning.

Aldebaran now lies to the west while the winter stars of Orion (Ori) can be seen rising over the south-eastern horizon. Look out for Betelgeuse and Rigel in the pre-dawn twilight.

This image depicts the sky looking south-east at about 75 minutes before sunrise.

September 11th to 20th, 2020

The Moon

| 11th | 13th | 15th | 17th | 19th |

Date	Con	R.A.	Dec	Mag	Diam	Ill.	Elon.	Phase	Close To
11th	Gem	6h 8m	24° 1'	-9.6	31'	39%	78° W	LQ	
12th	Gem	7h 4m	24° 18'	-9.1	31'	29%	65° W	-Cr	
13th	Cnc	8h 2m	23° 14'	-8.4	32'	20%	51° W	-Cr	Venus, Praesepe
14th	Cnc	9h 1m	20° 46'	-7.6	32'	12%	37° W	NM	Venus, Praesepe
15th	Leo	9h 59m	16° 59'	-6.6	33'	6%	24° W	NM	Regulus
16th	Leo	10h 56m	12° 4'	-5.4	33'	2%	11° W	NM	
17th	Vir	11h 51m	6° 19'	-4.4	33'	0%	3° E	NM	
18th	Vir	12h 47m	0° 9'	-5.5	33'	2%	15° E	NM	Mercury, Spica
19th	Vir	13h 42m	-6° 2'	-6.7	33'	6%	28° E	NM	Mercury, Spica
20th	Lib	14h 38m	-11° 48'	-7.8	33'	13%	41° E	+Cr	

Mercury and Venus

Mercury
15th

Venus
15th

Mercury

Date	Con.	R.A.	Dec.	Mag.	Diam.	Ill.	Elon.	Vis.	Rat.	Close To
11th	Vir	12h 31m	-3° 50'	-0.2	5"	84%	18° E	PM	**	
13th	Vir	12h 42m	-5° 14'	-0.2	5"	82%	19° E	PM	**	
15th	Vir	12h 52m	-6° 36'	-0.1	6"	80%	20° E	PM	**	Spica
17th	Vir	13h 2m	-7° 56'	-0.1	6"	78%	20° E	PM	***	Spica
19th	Vir	13h 12m	-9° 13'	-0.1	6"	76%	21° E	PM	***	Moon, Spica

Venus

Date	Con.	R.A.	Dec.	Mag.	Diam.	Ill.	Elon.	Vis.	Rat.	Close To
11th	Cnc	8h 31m	17° 47'	-4.2	18"	64%	42° W	AM	***	Praesepe
13th	Cnc	8h 40m	17° 22'	-4.1	18"	65%	42° W	AM	***	Moon, Praesepe
15th	Cnc	8h 49m	16° 56'	-4.1	17"	66%	41° W	AM	***	Praesepe
17th	Cnc	8h 58m	16° 27'	-4.1	17"	67%	41° W	AM	***	Praesepe
19th	Cnc	9h 7m	15° 57'	-4.1	17"	67%	40° W	AM	***	Praesepe

Mars and the Outer Planets

Mars
15th

Jupiter
15th

Saturn
15th

Mars

Date	Con.	R.A.	Dec.	Mag.	Diam.	Ill.	Elon.	Vis.	Rat.	Close To
11th	Psc	1h 50m	6° 52'	-2.1	20"	94%	142° W	AM	****	
15th	Psc	1h 49m	6° 50'	-2.2	21"	96%	146° W	AM	*****	
20th	Psc	1h 47m	6° 43'	-2.3	22"	97%	151° W	AM	*****	

The Outer Planets

Planet	Date	Con.	R.A.	Dec.	Mag.	Diam.	Elon.	Vis.	Rat.	Close To
Jupiter	15th	Sgr	19h 15m	-22° 43'	-2.5	43"	115° E	PM	****	Saturn
Saturn	15th	Sgr	19h 50m	-21° 20'	0.4	18"	124° E	PM	****	
Uranus	15th	Ari	2h 32m	14° 28'	5.7	4"	136° W	AM	***	
Neptune	15th	Aqr	23h 22m	-5° 12'	7.8	2"	177° E	AN	*****	

Highlights

Date	Time (UT)	Event
12th	09:20	Neptune is at opposition. (Visible all night.)
	23:20	Jupiter is stationary prior to resuming prograde motion. (Evening sky.)
13th	10:21	Venus is 2.3° south of the Praesepe star cluster. (Cancer, morning sky.)
14th	02:05	The waning crescent Moon is north of the Praesepe star cluster. (Cancer, morning sky.)
	03:24	The waning crescent Moon is north of Venus. (Morning sky.)
	N/A	Good opportunity to see Earthshine on the waning crescent Moon. (Morning sky.)
15th	17:03	The waning crescent Moon is north of the bright star Regulus. (Leo, morning sky.)
17th	11:01	New Moon. (Not visible.)
18th	22:48	The waxing crescent Moon is north of Mercury. (Evening sky.)
19th	03:55	The waxing crescent Moon is north of the bright star Spica. (Evening sky.)
20th	N/A	Good opportunity to see Earthshine on the waxing crescent Moon. (Evening sky.)

Planet Locations – September 15th

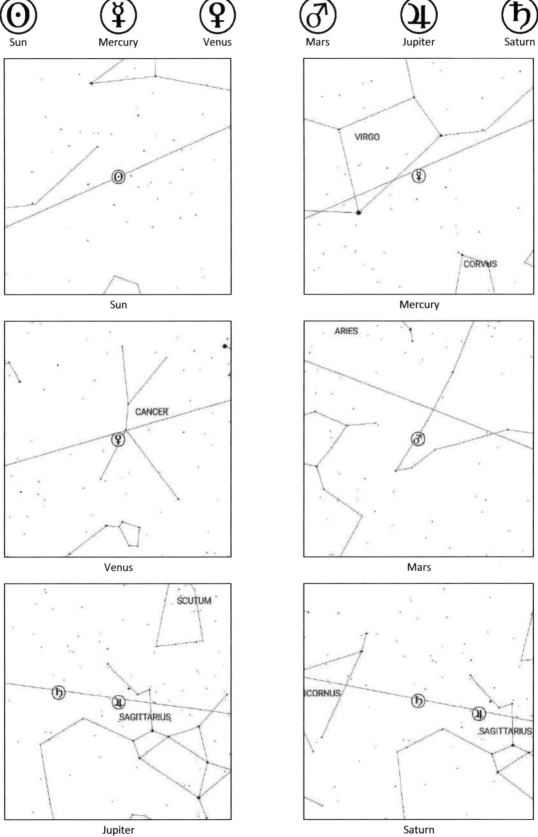

| Sun | Mercury | Venus | Mars | Jupiter | Saturn |

Sun

Mercury

Venus

Mars

Jupiter

Saturn

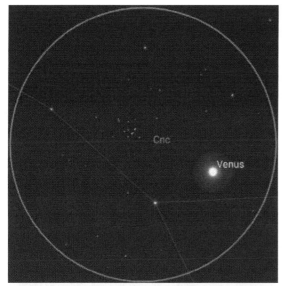

September 13th – Morning Sky

It's been a good year for observing Venus, with the planet making a number of close passes of other planets and star clusters.

This morning you can catch it within the same binocular field of view as the Praesepe, or Beehive star cluster. If you rise early enough, you should be able to see them both in the pre-dawn sky.

This image depicts the view through 10x50 binoculars looking east at about 90 minutes before sunrise.

September 14th – Morning Sky

If you missed Venus passing the Praesepe yesterday, you'll get another chance this morning. In fact, the pair will appear within the same binocular field of view until the 17th.

This morning they're joined by the waning crescent Moon. Be sure to look for Earthshine on the Moon's darkened surface.

This image depicts the sky looking east at about 45 minutes before sunrise.

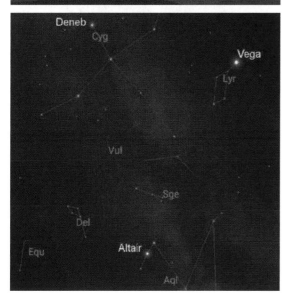

September 15th – Evening Sky

Three stars light up the southern sky during the late summer and early autumn months. These three stars form the Summer Triangle and are easily visible from almost anywhere in the northern hemisphere.

At less than 30 light years away, both Vega and Altair are celestial neighbors to our Sun. Deneb, however, is over 1,000 light years away. In order to appear this bright from so far away, it must shine with a luminosity of about 50,000 Suns!

This image depicts the sky looking south at about two hours after sunset.

September 21st to 30th, 2020

The Moon

21st

23rd

25th

27th

29th

Date	Con	R.A.	Dec	Mag	Diam	Ill.	Elon.	Phase	Close To
21st	Lib	15h 34m	-16° 48'	-8.6	33'	22%	55° E	+Cr	
22nd	Oph	16h 33m	-20° 41'	-9.3	32'	33%	68° E	+Cr	Antares
23rd	Oph	17h 32m	-23° 15'	-9.8	32'	44%	82° E	FQ	
24th	Sgr	18h 31m	-24° 24'	-10.3	31'	55%	96° E	FQ	Jupiter
25th	Sgr	19h 29m	-24° 7'	-10.7	31'	65%	110° E	FQ	Jupiter, Saturn
26th	Cap	20h 24m	-22° 34'	-11.1	31'	74%	123° E	+G	Saturn
27th	Cap	21h 17m	-19° 54'	-11.4	30'	83%	135° E	+G	
28th	Aqr	22h 7m	-16° 21'	-11.7	30'	89%	147° E	+G	
29th	Aqr	22h 55m	-12° 10'	-12.0	30'	95%	158° E	+G	Neptune
30th	Aqr	23h 40m	-7° 33'	-12.3	30'	98%	168° E	FM	Neptune

Mercury and Venus

Mercury
25th

Venus
25th

Mercury

Date	Con.	R.A.	Dec.	Mag.	Diam.	Ill.	Elon.	Vis.	Rat.	Close To
21st	Vir	13h 22m	-10° 27'	0.0	6"	74%	22° E	PM	***	Spica
23rd	Vir	13h 31m	-11° 37'	0.0	6"	72%	22° E	PM	***	Spica
25th	Vir	13h 40m	-12° 44'	0.0	6"	69%	22° E	PM	***	Spica
27th	Vir	13h 48m	-13° 47'	0.0	6"	67%	23° E	PM	***	Spica
29th	Vir	13h 57m	-14° 46'	0.0	7"	64%	23° E	PM	***	Spica

Venus

Date	Con.	R.A.	Dec.	Mag.	Diam.	Ill.	Elon.	Vis.	Rat.	Close To
21st	Cnc	9h 16m	15° 25'	-4.1	17"	68%	40° W	AM	**	Praesepe
23rd	Leo	9h 26m	14° 51'	-4.1	16"	69%	39° W	AM	**	
25th	Leo	9h 35m	14° 15'	-4.1	16"	70%	39° W	AM	**	Regulus
27th	Leo	9h 44m	13° 38'	-4.1	16"	70%	38° W	AM	**	Regulus
29th	Leo	9h 53m	12° 59'	-4.1	16"	71%	38° W	AM	**	Regulus

Mars and the Outer Planets

Mars
25th

Jupiter
25th

Saturn
25th

Mars

Date	Con.	R.A.	Dec.	Mag.	Diam.	Ill.	Elon.	Vis.	Rat.	Close To
21st	Psc	1h 46m	6° 41'	-2.3	22"	97%	152° W	AM	*****	
25th	Psc	1h 43m	6° 31'	-2.4	22"	98%	157° W	AM	*****	
30th	Psc	1h 38m	6° 16'	-2.5	22"	99%	162° W	AM	*****	

The Outer Planets

Planet	Date	Con.	R.A.	Dec.	Mag.	Diam.	Elon.	Vis.	Rat.	Close To
Jupiter	25th	Sgr	19h 16m	-22° 42'	-2.4	41"	107° E	PM	***	Moon, Saturn
Saturn	25th	Sgr	19h 49m	-21° 22'	0.4	17"	115° E	PM	***	Moon
Uranus	25th	Ari	2h 31m	14° 23'	5.7	4"	145° W	AM	***	
Neptune	25th	Aqr	23h 21m	-5° 19'	7.8	2"	168° E	PM	*****	

Highlights

Date	Time (UT)	Event
22nd	03:11	Mercury is 0.3° north of the bright star Spica. (Virgo, evening sky.)
	09:17	The waxing crescent Moon is north of the bright star Antares. (Scorpius, evening sky.)
	13:32	Autumn Equinox.
24th	01:56	First Quarter Moon. (Evening sky.)
25th	06:49	The just-past first quarter Moon is south of Jupiter. (Evening sky.)
	13:57	The just-past first quarter Moon is south of dwarf planet Pluto. (Evening sky.)
	21:08	The just-past first quarter Moon is south of Saturn. (Evening sky.)
29th	01:06	Saturn is stationary prior to resuming prograde motion. (Evening sky.)
30th	03:00	The nearly full Moon is south of Neptune. (Evening sky.)

Planet Locations – September 25th

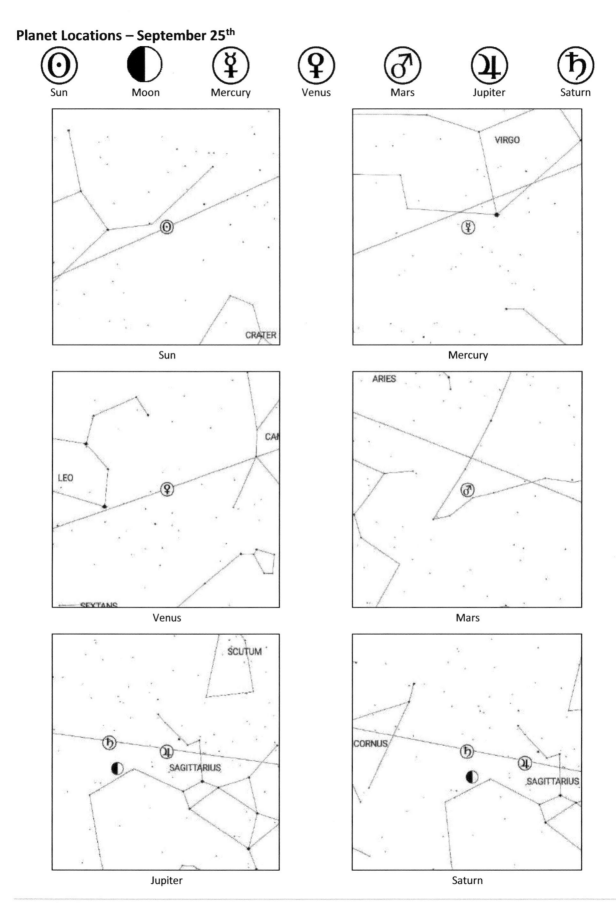

Sun   Moon   Mercury   Venus   Mars   Jupiter   Saturn

Sun

Mercury

Venus

Mars

Jupiter

Saturn

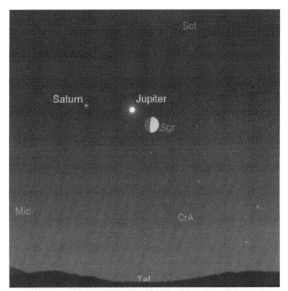

September 24th – Evening Sky

The Moon turned new about a week ago and has now reached first quarter. It's passing by the giant worlds of Jupiter and Saturn over the next few nights.

This evening you can see it to the lower right of Jupiter in the twilight after sunset. Come back again tomorrow to see it on the opposite side of Saturn.

This image depicts the sky looking south at about 45 minutes after sunset.

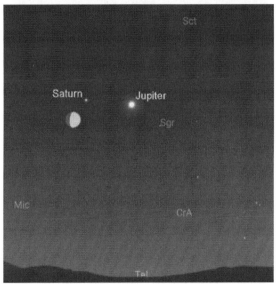

September 25th - Evening Sky

Yesterday the Moon appeared close to Jupiter, but tonight it's skipped past the planet and is now close to Saturn.

It's now a waxing gibbous and will turn full on October 1st.

This image depicts the sky looking south at about 45 minutes after sunset.

September 25th - Evening Sky

Autumn is when the major players of a famous Greek myth take to the skies. According to legend, Cassiopeia (Cas) the Queen once boasted of her beauty and offended the god Poseidon. He sent a monster to destroy her country, forcing her to offer her daughter Andromeda (And) as a sacrifice to appease the god.

Fortunately, in true fairytale fashion, the hero Perseus (Per) arrived in the nick of time to slay the monster and save the princess!

This image depicts the sky looking north-east at about two hours after sunset.

October 1ˢᵗ to 10ᵗʰ, 2020

The Moon

1ˢᵗ

3ʳᵈ

5ᵗʰ

7ᵗʰ

9ᵗʰ

Date	Con	R.A.	Dec	Mag	Diam	Ill.	Elon.	Phase	Close To
1st	Psc	0h 24m	-2° 40'	-12.6	30'	100%	178° E	FM	
2nd	Cet	1h 8m	2° 16'	-12.5	29'	99%	172° W	FM	Mars
3rd	Psc	1h 51m	7° 6'	-12.3	29'	98%	162° W	FM	Mars, Uranus
4th	Ari	2h 35m	11° 40'	-12.0	29'	94%	152° W	-G	Uranus
5th	Ari	3h 21m	15° 49'	-11.7	30'	89%	141° W	-G	Pleiades
6th	Tau	4h 9m	19° 22'	-11.4	30'	82%	130° W	-G	Pleiades, Hyades, Aldebaran
7th	Tau	4h 59m	22° 8'	-11.1	30'	74%	119° W	-G	Hyades, Aldebaran
8th	Tau	5h 51m	23° 57'	-10.7	30'	65%	107° W	LQ	
9th	Gem	6h 46m	24° 37'	-10.4	31'	55%	94° W	LQ	
10th	Gem	7h 42m	24° 1'	-9.9	31'	45%	81° W	LQ	

Mercury and Venus

Mercury
5ᵗʰ

Venus
5ᵗʰ

Mercury

Date	Con.	R.A.	Dec.	Mag.	Diam.	Ill.	Elon.	Vis.	Rat.	Close To
1st	Vir	14h 4m	-15° 40'	0.1	7"	60%	23° E	PM	***	Spica
3rd	Vir	14h 11m	-16° 28'	0.1	7"	57%	23° E	PM	***	
5th	Vir	14h 18m	-17° 10'	0.1	7"	53%	23° E	PM	***	
7th	Lib	14h 23m	-17° 44'	0.2	8"	48%	22° E	PM	***	
9th	Lib	14h 27m	-18° 10'	0.3	8"	43%	22° E	PM	***	

Venus

Date	Con.	R.A.	Dec.	Mag.	Diam.	Ill.	Elon.	Vis.	Rat.	Close To
1st	Leo	10h 2m	12° 18'	-4.1	16"	72%	37° W	AM	**	Regulus
3rd	Leo	10h 11m	11° 36'	-4.1	15"	72%	37° W	AM	**	Regulus
5th	Leo	10h 20m	10° 52'	-4.1	15"	73%	36° W	AM	**	Regulus
7th	Leo	10h 29m	10° 7'	-4.1	15"	74%	36° W	AM	**	Regulus
9th	Leo	10h 38m	9° 21'	-4.1	15"	75%	36° W	AM	**	Regulus

Mars and the Outer Planets

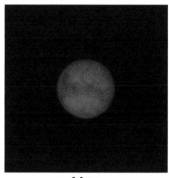

Mars	Jupiter	Saturn
5th	5th	5th

Mars

Date	Con.	R.A.	Dec.	Mag.	Diam.	Ill.	Elon.	Vis.	Rat.	Close To
1st	Psc	1h 37m	6° 13'	-2.5	22"	99%	164° W	AM	*****	
5th	Psc	1h 33m	5° 58'	-2.6	23"	99%	168° W	AM	*****	
10th	Psc	1h 26m	5° 39'	-2.6	23"	100%	175° W	AN	*****	

The Outer Planets

Planet	Date	Con.	R.A.	Dec.	Mag.	Diam.	Elon.	Vis.	Rat.	Close To
Jupiter	5th	Sgr	19h 19m	-22° 38'	-2.3	40"	98° E	PM	***	Saturn
Saturn	5th	Sgr	19h 49m	-21° 22'	0.5	17"	106° E	PM	***	
Uranus	5th	Ari	2h 29m	14° 16'	5.7	4"	154° W	AM	***	
Neptune	5th	Aqr	23h 21m	-5° 25'	7.8	2"	159° E	PM	*****	

Highlights

Date	Time (UT)	Event
1st	16:00	Mercury is at greatest eastern elongation from the Sun. (Evening sky.)
	21:06	Full Moon. (Visible all night.)
2nd	17:50	Venus is 0.1° south of the bright star Regulus. (Leo, morning sky.)
3rd	04:36	The waning gibbous Moon is south of Mars. (Morning sky.)
	19:02	Dwarf planet Pluto is stationary prior to resuming prograde motion. (Evening sky.)
4th	10:02	The waning gibbous Moon is south of Uranus. (Morning sky.)
6th	00:09	The waning gibbous Moon is south of the Pleiades star cluster. (Taurus, morning sky.)
	23:56	The waning gibbous Moon is north of the bright star Aldebaran. (Taurus, morning sky.)
8th	N/A	The Draconid meteor shower is at its maximum. (ZHR: Variable.)
10th	00:40	Last Quarter Moon. (Morning sky.)

Planet Locations – October 5th

Sun | Mercury | Venus | Mars | Jupiter | Saturn

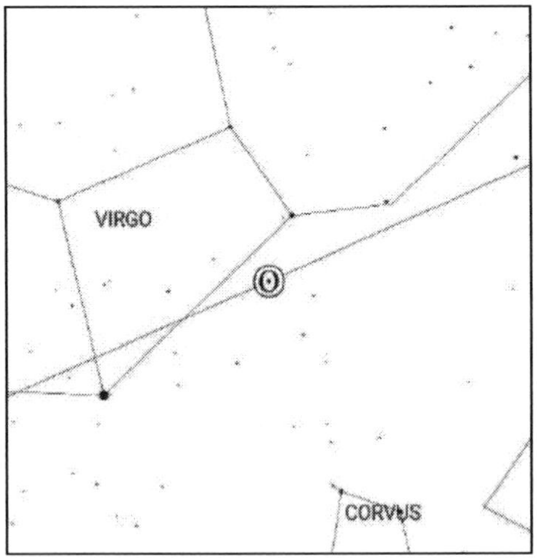

Sun

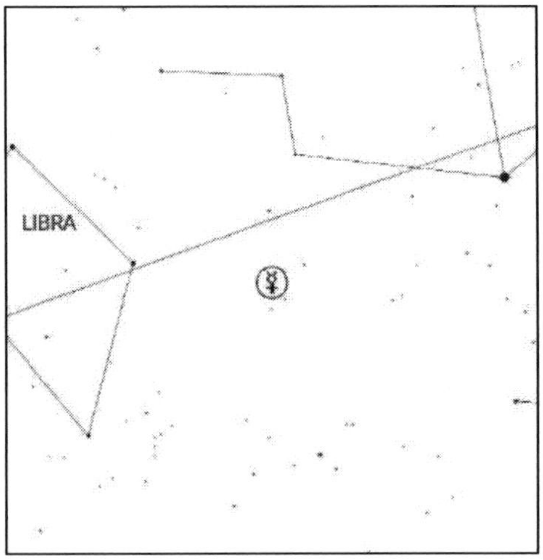

Mercury

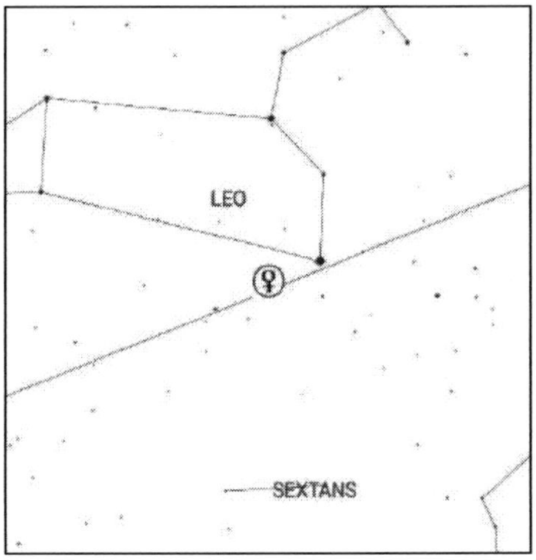

Venus

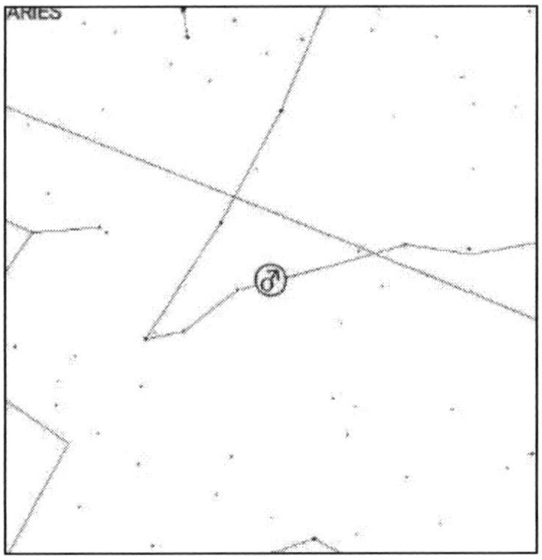

Mars

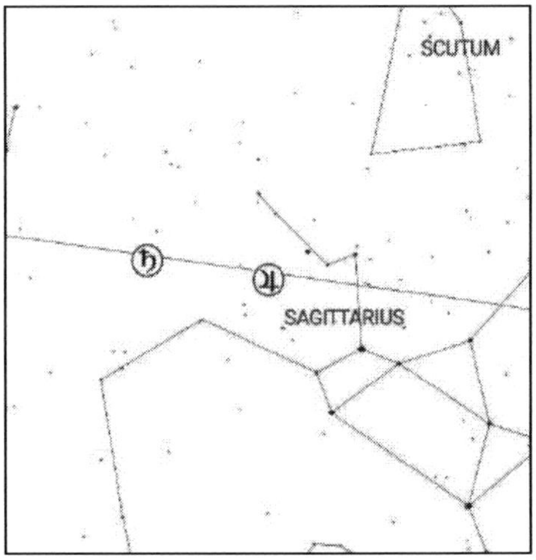

Jupiter

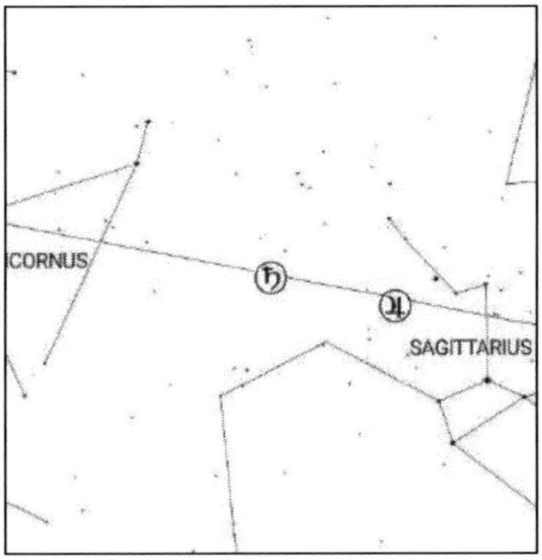

Saturn

October 1st – Evening Sky

Mercury is at greatest eastern elongation from the Sun today and the planet may be visible in the evening twilight just after sunset.

You'll need a clear, unobstructed view of the south-western horizon as the planet will be quite low and tricky to spot. If you don't see it tonight, you'll have about another two weeks, but neither the Moon or other planets appear nearby to help you find it.

This image depicts the sky looking south-west at about 30 minutes after sunset.

October 2nd – Morning Sky

If you've enjoyed seeing Venus close to other planets, stars and star clusters this year, you'll want to get up early today.

This morning it appears very close to Regulus, the brightest star in the constellation of Leo, the Lion. Both can easily be seen together within the same binocular field of view.

This image depicts the view through 10x50 binoculars, looking east at about 45 minutes before sunrise.

October 3rd – Morning Sky

Mars is nearly at opposition and is currently at its brightest. It outshines everything else in the sky except the Sun and Moon.

You can see the waning gibbous Moon to the upper left of the planet in the pre-dawn sky this morning.

This image depicts the sky looking west at about 45 minutes before sunrise.

October 11th to 20th, 2020

The Moon

| 11th | 13th | 15th | 17th | 19th |

Date	Con	R.A.	Dec	Mag	Diam	Ill.	Elon.	Phase	Close To
11th	Cnc	8h 39m	22° 6'	-9.4	31'	35%	67° W	-Cr	Praesepe
12th	Leo	9h 35m	18° 52'	-8.8	32'	25%	54° W	-Cr	Regulus
13th	Leo	10h 31m	14° 28'	-8.0	32'	16%	41° W	-Cr	Venus, Regulus
14th	Leo	11h 26m	9° 4'	-7.0	33'	8%	28° W	NM	Venus
15th	Vir	12h 21m	3° 2'	-5.9	33'	3%	15° W	NM	
16th	Vir	13h 17m	-3° 18'	-4.6	33'	0%	2° W	NM	Spica
17th	Vir	14h 13m	-9° 28'	-5.0	33'	1%	11° E	NM	Mercury
18th	Lib	15h 11m	-15° 2'	-6.3	33'	4%	24° E	NM	Mercury
19th	Sco	16h 11m	-19° 35'	-7.4	33'	11%	38° E	NM	Antares
20th	Oph	17h 12m	-22° 46'	-8.3	32'	19%	52° E	+Cr	Antares

Mercury and Venus

Mercury
15th

Venus
15th

Mercury

Date	Con.	R.A.	Dec.	Mag.	Diam.	Ill.	Elon.	Vis.	Rat.	Close To
11th	Lib	14h 31m	-18° 26'	0.4	8"	38%	21° E	PM	***	
13th	Lib	14h 32m	-18° 31'	0.6	9"	31%	19° E	PM	***	
15th	Lib	14h 32m	-18° 21'	0.9	9"	25%	17° E	PM	***	
17th	Lib	14h 29m	-17° 57'	1.4	9"	18%	15° E	NV	N/A	Moon
19th	Lib	14h 25m	-17° 14'	2.0	10"	12%	12° E	NV	N/A	

Venus

Date	Con.	R.A.	Dec.	Mag.	Diam.	Ill.	Elon.	Vis.	Rat.	Close To
11th	Leo	10h 47m	8° 33'	-4.1	15"	75%	35° W	AM	**	Regulus
13th	Leo	10h 56m	7° 44'	-4.1	14"	76%	35° W	AM	**	Moon
15th	Leo	11h 5m	6° 55'	-4.0	14"	76%	34° W	AM	**	
17th	Leo	11h 14m	6° 4'	-4.0	14"	77%	34° W	AM	**	
19th	Leo	11h 23m	5° 12'	-4.0	14"	78%	34° W	AM	**	

Mars and the Outer Planets

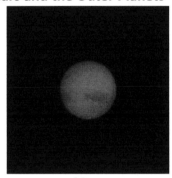  

Mars	Jupiter	Saturn
15th	15th	15th

Mars

Date	Con.	R.A.	Dec.	Mag.	Diam.	Ill.	Elon.	Vis.	Rat.	Close To
11th	Psc	1h 25m	5° 36'	-2.6	22"	100%	176° W	AN	*****	
15th	Psc	1h 20m	5° 21'	-2.6	22"	100%	179° E	AN	*****	
20th	Psc	1h 14m	5° 6'	-2.5	22"	100%	173° E	AN	*****	

The Outer Planets

Planet	Date	Con.	R.A.	Dec.	Mag.	Diam.	Elon.	Vis.	Rat.	Close To
Jupiter	15th	Sgr	19h 22m	-22° 31'	-2.3	39"	90° E	PM	***	Saturn
Saturn	15th	Sgr	19h 50m	-21° 21'	0.5	17"	97° E	PM	***	
Uranus	15th	Ari	2h 28m	14° 9'	5.7	4"	164° W	AM	****	
Neptune	15th	Aqr	23h 20m	-5° 30'	7.8	2"	149° E	PM	****	

Highlights

Date	Time (UT)	Event
11th	13:39	The just-past last quarter Moon is north of the Praesepe star cluster. (Cancer, morning sky.)
13th	00:52	The waning crescent Moon is north of the bright star Regulus. (Leo, morning sky.)
	23:18	The waning crescent Moon is north of Venus. (Morning sky.)
	N/A	Good opportunity to see Earthshine on the waning crescent Moon. (Morning sky.)
14th	04:29	Mercury is stationary prior to beginning retrograde motion. (Evening sky.)
	21:17	Mars is at opposition. (Visible all night.)
16th	19:32	New Moon. (Not visible.)
18th	N/A	The Epsilon Geminid meteor shower is at its maximum. (ZHR: 3)
19th	20:28	The waxing crescent Moon is north of the bright star Antares. (Scorpius, evening sky.)
20th	N/A	Good opportunity to see Earthshine on the waxing crescent Moon. (Evening sky.)

Planet Locations – October 15th

| Sun | Mercury | Venus | | Mars | Jupiter | Saturn |

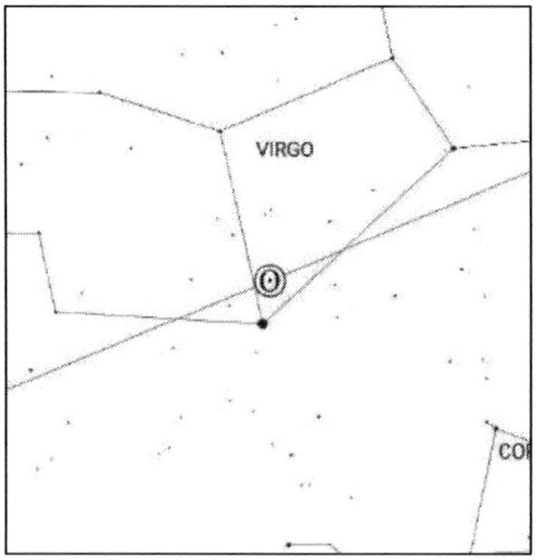

Sun

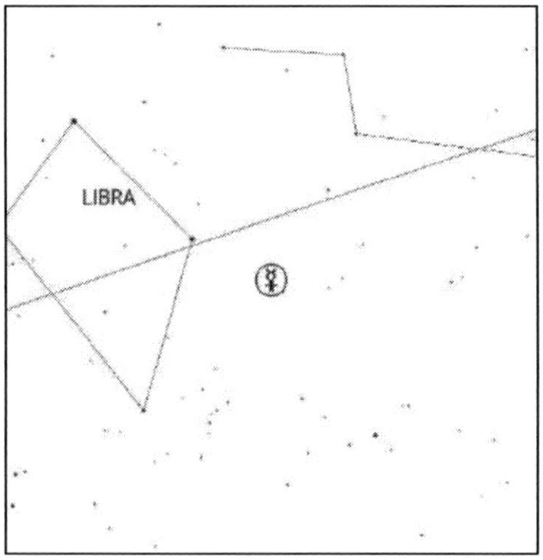

Mercury

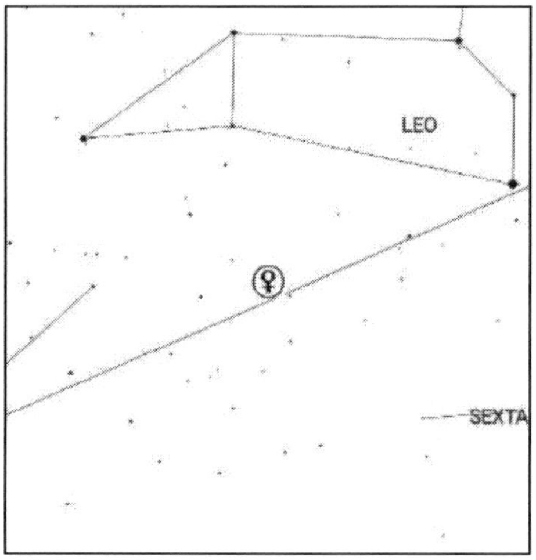

Venus

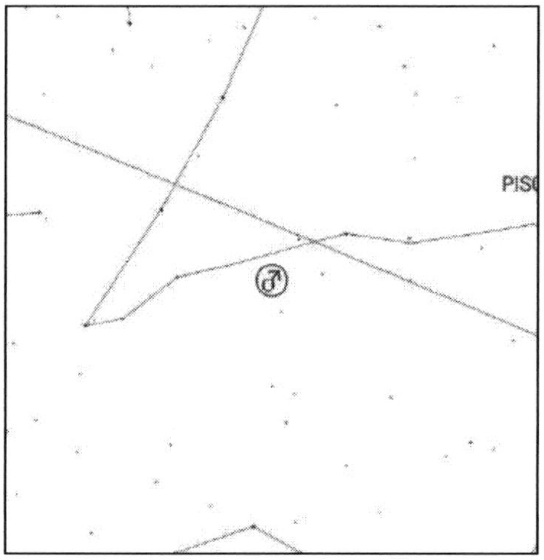

Mars

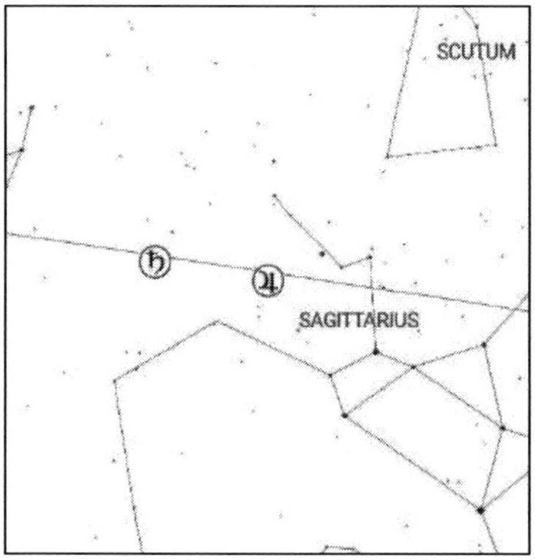

Jupiter

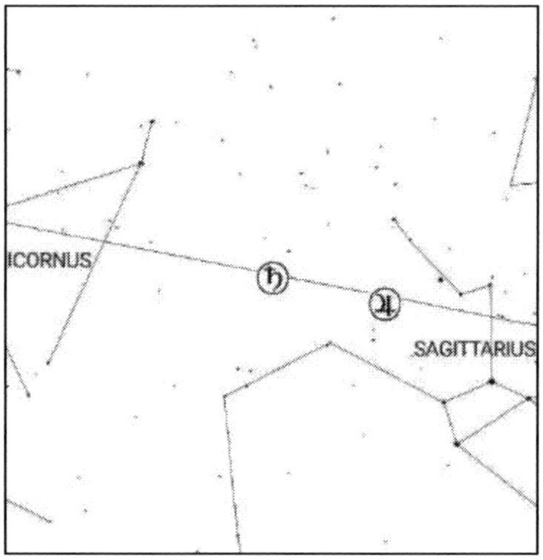

Saturn

October 13th – Morning Sky

After passing Mars ten days ago, the Moon is now approaching Venus. It's slimmed down to a waning crescent and can be seen almost directly above our nearest planetary neighbor.

It will appear on the other side of the planet tomorrow morning. In the meantime, take a moment to enjoy the Earthshine illuminating the Moon's darkened surface.

This image depicts the sky looking east about an hour before sunrise.

October 14th – Evening Sky

Mars is at opposition, which means it's visible throughout the night. It will rise at sunset and then set at sunrise.

It's an unmissable sight; nothing else comes close to its brightness or its color. The planet won't be this bright again until 2035. Most people describe its light as appearing coppery – what color do you see?

This image depicts the sky looking east at about an hour and 45 minutes after sunset.

October 19th – Evening Sky

There was a new Moon three days ago and now the Moon pays a final visit to the bright star Antares. You can catch the pair low over the south-western horizon in the evening twilight.

Meanwhile, Jupiter and Saturn await their turn toward the south. They'll be visited by the Moon on the 22nd.

This image depicts the sky looking south-west at about an hour after sunset.

October 21ˢᵗ to 31ˢᵗ, 2020

The Moon

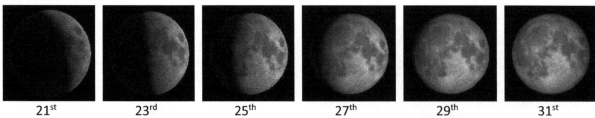

| 21st | 23rd | 25th | 27th | 29th | 31st |

Date	Con	R.A.	Dec	Mag	Diam	Ill.	Elon.	Phase	Close To
21st	Sgr	18h 13m	-24° 27'	-9.0	32'	28%	67° E	+Cr	
22nd	Sgr	19h 13m	-24° 35'	-9.6	31'	39%	81° E	FQ	Jupiter, Saturn
23rd	Cap	20h 11m	-23° 19'	-10.1	31'	49%	94° E	FQ	Jupiter, Saturn
24th	Cap	21h 5m	-20° 51'	-10.5	31'	60%	107° E	FQ	
25th	Cap	21h 56m	-17° 27'	-10.9	30'	69%	119° E	+G	
26th	Aqr	22h 44m	-13° 22'	-11.2	30'	78%	130° E	+G	Neptune
27th	Aqr	23h 29m	-8° 49'	-11.5	30'	85%	140° E	+G	Neptune
28th	Psc	0h 13m	-3° 59'	-11.8	30'	91%	150° E	+G	
29th	Cet	0h 57m	0° 59'	-12.1	29'	96%	160° E	FM	Mars
30th	Psc	1h 40m	5° 53'	-12.4	29'	99%	170° E	FM	Mars, Uranus
31st	Ari	2h 24m	10° 35'	-12.6	29'	100%	180° E	FM	Uranus

Mercury and Venus

Mercury
25ᵗʰ

Venus
25ᵗʰ

Mercury

Date	Con.	R.A.	Dec.	Mag.	Diam.	Ill.	Elon.	Vis.	Rat.	Close To
21st	Vir	14h 19m	-16° 14'	2.9	10"	6%	8° E	NV	N/A	
23rd	Vir	14h 11m	-14° 58'	4.0	10"	2%	4° E	NV	N/A	
25th	Vir	14h 2m	-13° 31'	5.5	10"	0%	0° E	NV	N/A	Spica
27th	Vir	13h 53m	-12° 0'	4.4	10"	1%	4° W	NV	N/A	Spica
29th	Vir	13h 46m	-10° 37'	2.9	10"	5%	8° W	NV	N/A	Spica
31st	Vir	13h 41m	-9° 29'	1.8	9"	12%	11° W	NV	N/A	Spica

Venus

Date	Con.	R.A.	Dec.	Mag.	Diam.	Ill.	Elon.	Vis.	Rat.	Close To
21st	Leo	11h 32m	4° 20'	-4.0	14"	78%	33° W	AM	**	
23rd	Vir	11h 41m	3° 27'	-4.0	14"	79%	33° W	AM	**	
25th	Vir	11h 50m	2° 33'	-4.0	14"	80%	33° W	AM	**	
27th	Vir	11h 59m	1° 38'	-4.0	13"	80%	32° W	AM	**	
29th	Vir	12h 8m	0° 44'	-4.0	13"	81%	32° W	AM	**	
31st	Vir	12h 17m	0° 11'	-4.0	13"	81%	32° W	AM	**	

Mars and the Outer Planets

Mars	Jupiter	Saturn
25th	25th	25th

Mars

Date	Con.	R.A.	Dec.	Mag.	Diam.	Ill.	Elon.	Vis.	Rat.	Close To
21st	Psc	1h 13m	5° 4'	-2.5	22"	100%	172° E	AN	*****	
25th	Psc	1h 8m	4° 55'	-2.3	21"	99%	167° E	PM	*****	
31st	Psc	1h 3m	4° 49'	-2.2	20"	98%	160° E	PM	*****	

The Outer Planets

Planet	Date	Con.	R.A.	Dec.	Mag.	Diam.	Elon.	Vis.	Rat.	Close To
Jupiter	25th	Sgr	19h 27m	-22° 22'	-2.2	38"	82° E	PM	***	Saturn
Saturn	25th	Sgr	19h 51m	-21° 17'	0.6	17"	88° E	PM	***	
Uranus	25th	Ari	2h 26m	14° 2'	5.7	4"	174° W	AN	****	
Neptune	25th	Aqr	23h 19m	-5° 35'	7.8	2"	140° E	PM	****	

Highlights

Date	Time (UT)	Event
21st	N/A	The Orionid meteor shower is at its maximum. (ZHR: 25)
22nd	17:04	The nearly first quarter Moon is south of Jupiter. (Evening sky.)
	22:51	The nearly first quarter Moon is south of the dwarf planet Pluto. (Evening sky.)
23rd	04:08	The almost first quarter Moon is south of Saturn. (Evening sky.)
	13:24	First Quarter Moon. (Evening sky.)
24th	N/A	The Leo Minorid meteor shower is at its maximum. (ZHR: 2)
25th	18:17	Mercury is at inferior conjunction with the Sun. (Not visible.)
27th	06:58	The waxing gibbous Moon is south of Neptune. (Evening sky.)
29th	15:02	The waxing gibbous Moon is south of Mars. (Evening sky.)
31st	12:32	The almost full Moon is south of Uranus. (Visible all night.)
	14:50	Full Moon. (Visible all night.)
	17:15	Uranus is at opposition. (Visible all night.)

Planet Locations – October 25th

| Sun | Mercury | Venus | Mars | Jupiter | Saturn |

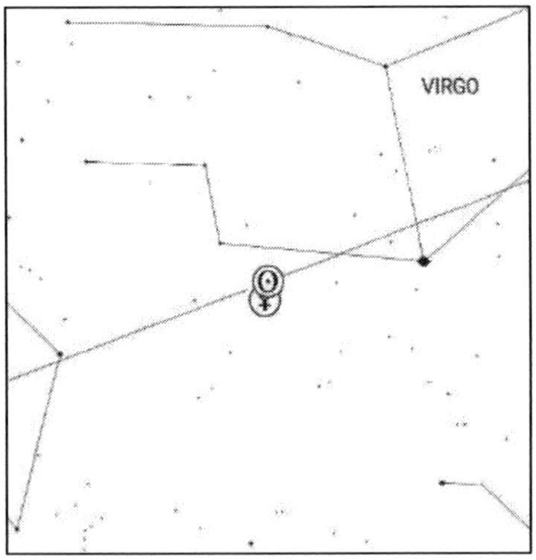

Sun

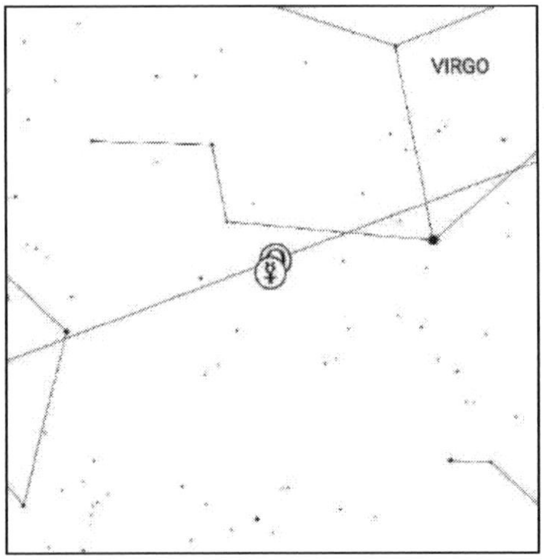

Mercury

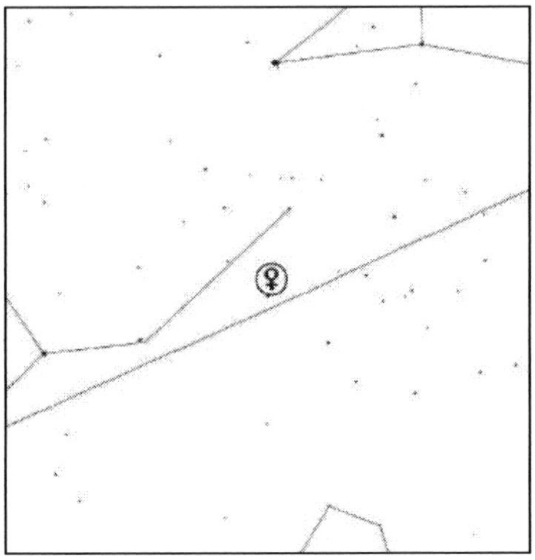

Venus

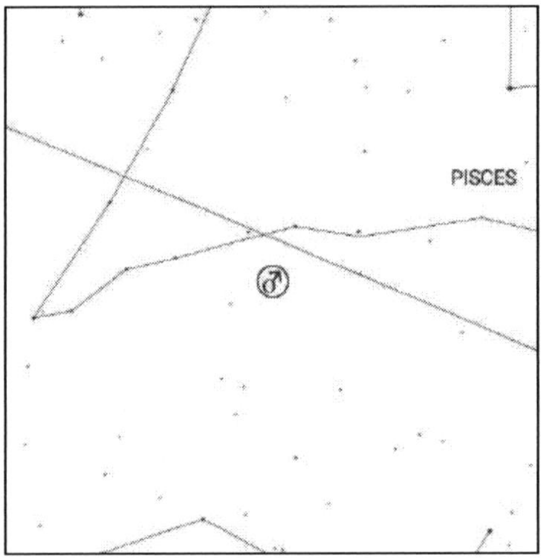

Mars

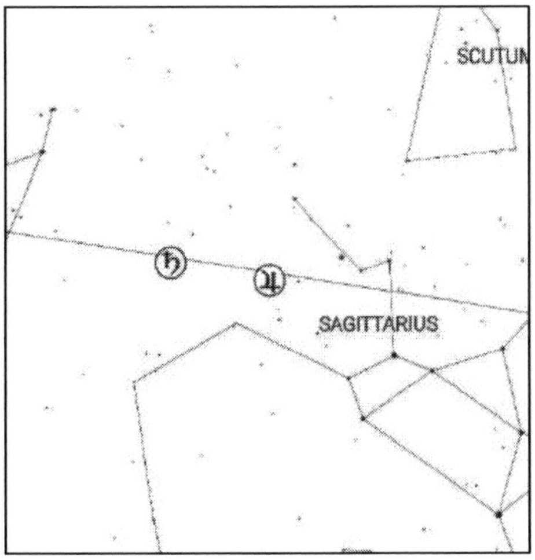

Jupiter

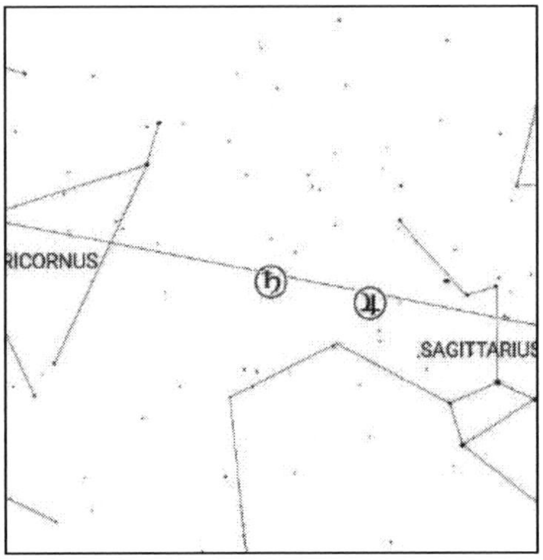

Saturn

October 20th to 21st – Morning & Evening Sky

The Orionid meteor shower reaches its maximum tonight. Fortunately, the waning crescent Moon won't provide much interference, especially during the later hours when the Orion (Ori) is rising over the horizon.

Under ideal conditions you could expect to see about 25 meteors an hour.

This image depicts the sky looking east at about 11:30 p.m.

October 22nd – Evening Sky

Just one day before reaching first quarter, the Moon appears between Jupiter and Saturn in the evening sky tonight.

Which planet appears closer to the Moon? You should be able to comfortably fit the Moon and one of the planets within the same 10x50 binocular view, but not both.

This image depicts the sky looking south-west at about two hours after sunset.

October 29th – Evening Sky

The Moon has returned to Mars, but whereas it was two days past full on the 3rd, it's two days before full tonight.

This means we'll be treated to a full Moon on Halloween, and with the blood-red Mars also shining brilliantly, it's sure to be a spooky night.

This image depicts the sky looking east at about two hours after sunset.

November 1st to 10th, 2020

The Moon

| 1st | 3rd | 5th | 7th | 9th |

Date	Con	R.A.	Dec	Mag	Diam	Ill.	Elon.	Phase	Close To
1st	Ari	3h 9m	14° 55'	-12.5	29'	99%	170° W	FM	Uranus, Pleiades
2nd	Tau	3h 56m	18° 40'	-12.2	30'	97%	159° W	FM	Pleiades, Hyades, Aldebaran
3rd	Tau	4h 46m	21° 41'	-11.9	30'	93%	147° W	-G	Hyades, Aldebaran
4th	Tau	5h 38m	23° 46'	-11.6	30'	87%	135° W	-G	
5th	Gem	6h 32m	24° 45'	-11.3	30'	80%	123° W	-G	
6th	Gem	7h 27m	24° 31'	-11.0	31'	71%	110° W	-G	
7th	Cnc	8h 22m	23° 0'	-10.6	31'	61%	97° W	LQ	Praesepe
8th	Cnc	9h 17m	20° 14'	-10.2	31'	51%	85° W	LQ	Praesepe
9th	Leo	10h 12m	16° 19'	-9.7	32'	40%	72° W	LQ	Regulus
10th	Leo	11h 5m	11° 25'	-9.1	32'	29%	60° W	-Cr	

Mercury and Venus

Mercury

5th

Venus
5th

Mercury

Date	Con.	R.A.	Dec.	Mag.	Diam.	Ill.	Elon.	Vis.	Rat.	Close To
1st	Vir	13h 39m	-9° 3'	1.3	9"	16%	12° W	NV	N/A	Spica
3rd	Vir	13h 38m	-8° 28'	0.6	8"	26%	14° W	NV	N/A	Spica
5th	Vir	13h 39m	-8° 17'	0.1	8"	35%	16° W	AM	**	Spica
7th	Vir	13h 43m	-8° 26'	-0.2	7"	45%	17° W	AM	***	Spica
9th	Vir	13h 49m	-8° 52'	-0.4	7"	54%	18° W	AM	***	Spica

Venus

Date	Con.	R.A.	Dec.	Mag.	Diam.	Ill.	Elon.	Vis.	Rat.	Close To
1st	Vir	12h 22m	0° 39'	-4.0	13"	82%	32° W	AM	**	
3rd	Vir	12h 31m	-1° 35'	-4.0	13"	82%	31° W	AM	**	
5th	Vir	12h 40m	-2° 30'	-4.0	13"	83%	31° W	AM	**	
7th	Vir	12h 49m	-3° 26'	-4.0	13"	83%	31° W	AM	**	Spica
9th	Vir	12h 58m	-4° 21'	-4.0	13"	84%	30° W	AM	**	Spica

Mars and the Outer Planets

Mars
5th

Jupiter
5th

Saturn
5th

Mars

Date	Con.	R.A.	Dec.	Mag.	Diam.	Ill.	Elon.	Vis.	Rat.	Close To
1st	Psc	1h 2m	4° 49'	-2.1	20"	98%	159° E	PM	*****	
5th	Psc	1h 0m	4° 51'	-2.0	19"	97%	154° E	PM	****	
10th	Psc	0h 58m	4° 60'	-1.8	18"	96%	148° E	PM	****	

The Outer Planets

Planet	Date	Con.	R.A.	Dec.	Mag.	Diam.	Elon.	Vis.	Rat.	Close To
Jupiter	5th	Sgr	19h 34m	-22° 8'	-2.1	37"	72° E	PM	**	Saturn
Saturn	5th	Sgr	19h 54m	-21° 12'	0.6	16"	77° E	PM	**	
Uranus	5th	Ari	2h 25m	13° 53'	5.7	4"	175° E	AN	****	
Neptune	5th	Aqr	23h 18m	-5° 38'	7.8	2"	129° E	PM	****	

Highlights

Date	Time (UT)	Event
2nd	08:17	The waning gibbous Moon is south of the Pleiades star cluster. (Taurus, morning sky.)
3rd	08:15	The waning gibbous Moon is north of the bright star Aldebaran. (Taurus, morning sky.)
	08:16	Mercury is stationary prior to resuming prograde motion. (Morning sky.)
7th	19:04	The nearly last quarter Moon is north of the Praesepe star cluster. (Morning sky.)
8th	13:47	Last Quarter Moon. (Morning sky.)
9th	11:34	The just-past last quarter Moon is north of the bright star Regulus. (Morning sky.)
10th	16:57	Mercury is at greatest western elongation from the Sun. (Morning sky.)

Planet Locations – November 5th

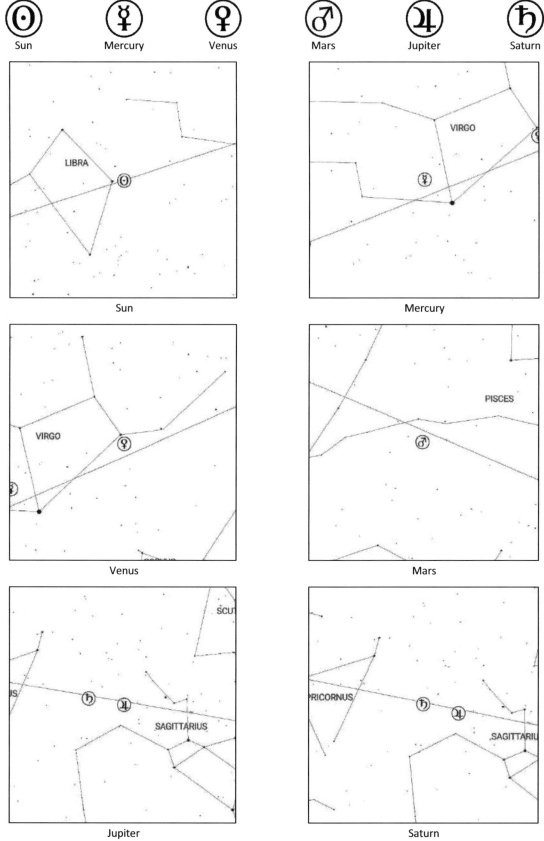

Sun

Mercury

Venus

Mars

Jupiter

Saturn

November 8th – Morning Sky

The Moon reaches last quarter this morning and is headed toward Regulus, the brightest star in Leo, the Lion.

It appears some way above the star this morning but that gap will close over the next twenty-four hours.

This image depicts the sky looking south-east at about 45 minutes before sunrise.

November 9th – Morning Sky

As promised, there's now a half Moon hanging to the upper left of Regulus this morning.

Yesterday there was 15 degrees between them, but this morning the distance has shrunk to just four. How many Moons do you think could fit between them?

This image depicts the sky looking south-east at about 45 minutes before sunrise.

November 10th – Morning Sky

Mercury reaches greatest western elongation from the Sun today and is fairly well placed for observation in the pre-dawn sky.

This is your last good chance to see the planet before the end of the year. It'll remain visible for about another ten days, but then it'll disappear until mid January.

Fortunately, Venus provides a convenient marker and can help your search this morning.

This image depicts the sky looking south-east at about 45 minutes before sunrise.

November 11th to 20th, 2020

The Moon

    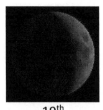

| 11th | 13th | 15th | 17th | 19th |

Date	Con	R.A.	Dec	Mag	Diam	Ill.	Elon.	Phase		Close To
11th	Vir	11h 58m	5° 45'	-8.3	33'	19%	47° W	-Cr		
12th	Vir	12h 52m	0° 23'	-7.5	33'	11%	35° W	NM		Venus, Spica
13th	Vir	13h 47m	-6° 37'	-6.4	33'	5%	22° W	NM		Mercury, Venus, Spica
14th	Lib	14h 44m	-12° 32'	-5.0	33'	1%	9° W	NM		Mercury
15th	Lib	15h 43m	-17° 41'	-4.4	33'	0%	5° E	NM		Antares
16th	Oph	16h 44m	-21° 38'	-5.8	33'	2%	19° E	NM		Antares
17th	Sgr	17h 47m	-24° 4'	-6.9	33'	7%	34° E	NM		
18th	Sgr	18h 50m	-24° 51'	-7.9	32'	15%	48° E	+Cr		
19th	Sgr	19h 51m	-24° 4'	-8.7	32'	23%	62° E	+Cr		Jupiter, Saturn
20th	Cap	20h 48m	-21° 55'	-9.3	31'	33%	76° E	+Cr		

Mercury and Venus

Mercury
15th

Venus
15th

Mercury

Date	Con.	R.A.	Dec.	Mag.	Diam.	Ill.	Elon.	Vis.	Rat.		Close To
11th	Vir	13h 57m	-9° 33'	-0.6	7"	62%	18° W	AM	***		Spica
13th	Vir	14h 5m	-10° 23'	-0.7	6"	69%	18° W	AM	**		Moon
15th	Vir	14h 15m	-11° 20'	-0.7	6"	74%	17° W	AM	**		
17th	Lib	14h 26m	-12° 21'	-0.7	6"	79%	17° W	AM	**		
19th	Lib	14h 37m	-13° 25'	-0.7	6"	83%	16° W	AM	**		

Venus

Date	Con.	R.A.	Dec.	Mag.	Diam.	Ill.	Elon.	Vis.	Rat.	Close To
11th	Vir	13h 7m	-5° 17'	-4.0	13"	84%	30° W	AM	**	Spica
13th	Vir	13h 16m	-6° 12'	-4.0	12"	85%	30° W	AM	**	Moon, Spica
15th	Vir	13h 25m	-7° 7'	-4.0	12"	85%	30° W	AM	**	Spica
17th	Vir	13h 35m	-8° 1'	-4.0	12"	86%	29° W	AM	**	Spica
19th	Vir	13h 44m	-8° 55'	-4.0	12"	86%	29° W	AM	**	Spica

Mars and the Outer Planets

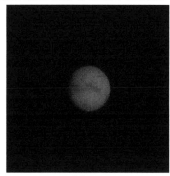
Mars
15th

Jupiter
15th

Saturn
15th

Mars

Date	Con.	R.A.	Dec.	Mag.	Diam.	Ill.	Elon.	Vis.	Rat.	Close To
11th	Psc	0h 57m	5° 2'	-1.8	18"	96%	147° E	PM	****	
15th	Psc	0h 57m	5° 15'	-1.7	17"	95%	143° E	PM	****	
20th	Psc	0h 57m	5° 36'	-1.5	16"	94%	138° E	PM	****	

The Outer Planets

Planet	Date	Con.	R.A.	Dec.	Mag.	Diam.	Elon.	Vis.	Rat.	Close To
Jupiter	15th	Sgr	19h 40m	-21° 52'	-2.1	36"	64° E	PM	**	Saturn
Saturn	15th	Sgr	19h 57m	-21° 4'	0.6	16"	68° E	PM	**	
Uranus	15th	Ari	2h 23m	13° 45'	5.7	4"	165° E	PM	****	
Neptune	15th	Aqr	23h 18m	-5° 40'	7.9	2"	118° E	PM	****	

Highlights

Date	Time (UT)	Event
12th	21:34	The waning crescent Moon is north of Venus. (Morning sky.)
	N/A	Good opportunity to see Earthshine on the waning crescent Moon. (Morning sky.)
	N/A	The Northern Taurid meteor shower is at its maximum. (ZHR: 5)
13th	01:24	The waning crescent Moon is north of the bright star Spica. (Virgo, morning sky.)
	21:14	The waning crescent Moon is north of Mercury. (Morning sky.)
15th	05:08	New Moon. (Not visible.)
	07:34	Venus is 4.1° north of the bright star Spica. (Virgo, morning sky.)
	19:42	Mars is stationary prior to resuming prograde motion. (Evening sky.)
18th	N/A	Good opportunity to see Earthshine on the waxing crescent Moon. (Evening sky.)
	N/A	The Leonid meteor shower is at its maximum. (ZHR: 20)
19th	06:14	The waxing crescent Moon is south of dwarf planet Pluto. (Evening sky.)
	07:52	The waxing crescent Moon is south of Jupiter. (Evening sky.)
	14:19	The waxing crescent Moon is south of Saturn. (Evening sky.)

Planet Locations – November 15th

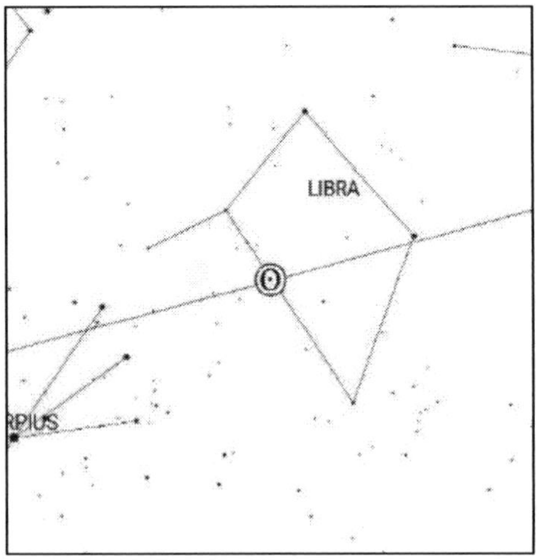

| ☉ Sun | ☿ Mercury | ♀ Venus | ♂ Mars | ♃ Jupiter | ♄ Saturn |

Sun

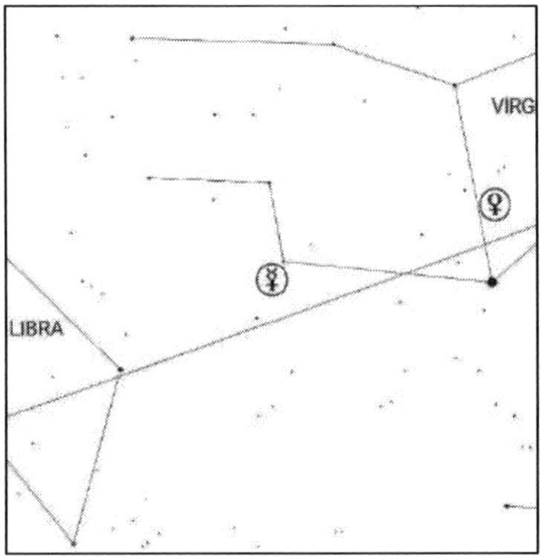

Mercury

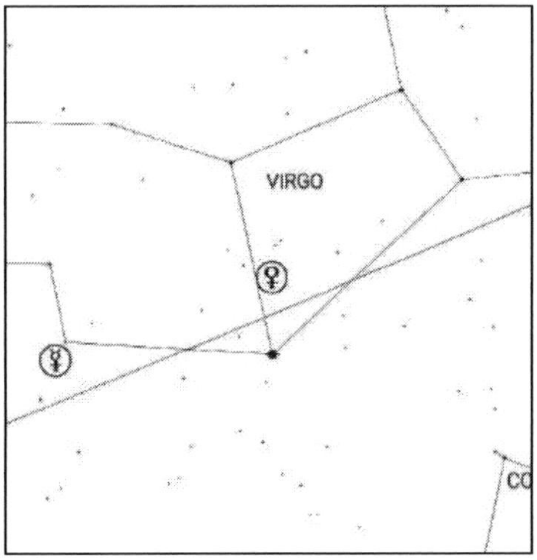

Venus

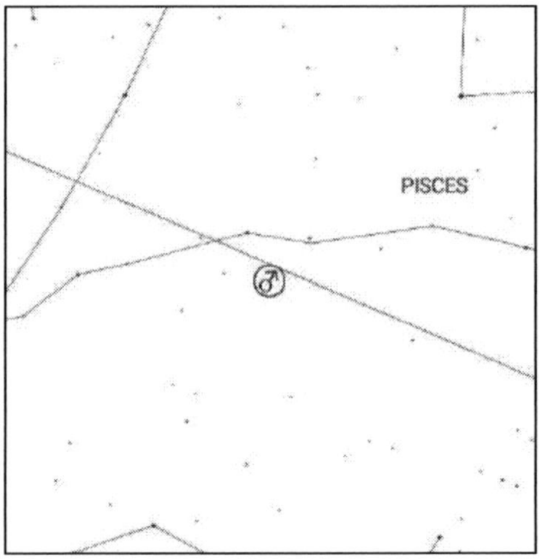

Mars

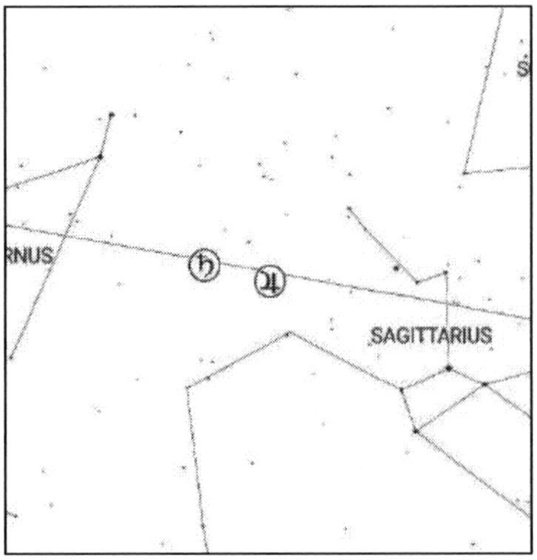

Jupiter

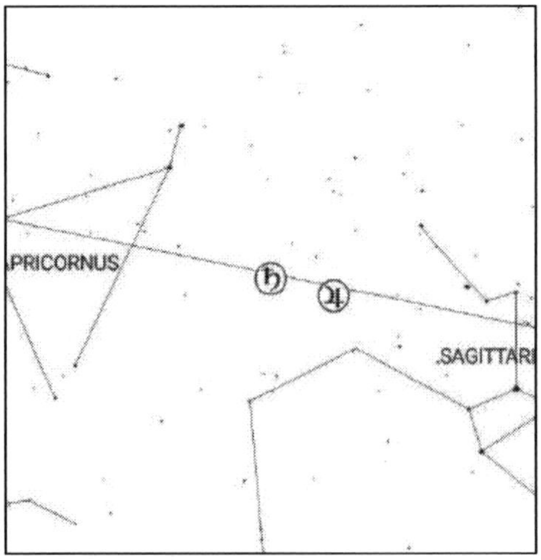

Saturn

November 12th – Morning Sky

If you've had no luck spotting Mercury over the past few days, try again this morning. The waning crescent Moon has entered the scene and can be seen directly above Venus.

Earthshine should also be easily visible on the lunar surface and look out too for Spica. The star may be seen twinkling some way to the lower right of Venus.

This image depicts the sky looking south-east at about 45 minutes before sunrise.

November 17th to 18th – Evening & Morning Sky

The Leonid meteor shower reaches its maximum today, and with the Moon turning new three days ago, this is a perfect time to catch some shooting stars.

Leo itself doesn't rise until the early hours, but you should still be able to see the meteors throughout the night. Under ideal conditions, you could see about 20 an hour.

This image depicts the sky looking east at about 1:00 a.m.

November 18th – Evening Sky

If you want to see the Moon for yourself, you can see the waxing crescent Moon in the twilight sky this evening.

It'll pass by the planets Jupiter and Saturn over the next few days, but will appear to the lower right of the pair tonight. Look out for Earthshine on the darkened lunar surface too.

This image depicts the sky looking south-west at about an hour after sunset.

November 21st to 30th, 2020

The Moon

21st

23rd

25th

27th

29th

Date	Con	R.A.	Dec	Mag	Diam	Ill.	Elon.	Phase		Close To
21st	Cap	21h 41m	-18° 42'	-9.8	31'	43%	88° E	FQ		
22nd	Aqr	22h 31m	-14° 43'	-10.3	30'	53%	99° E	FQ		Neptune
23rd	Aqr	23h 18m	-10° 12'	-10.6	30'	63%	110° E	FQ		Neptune
24th	Psc	0h 2m	-5° 23'	-11.0	30'	72%	120° E	+G		Neptune
25th	Cet	0h 45m	0° 26'	-11.3	30'	80%	130° E	+G		Mars
26th	Psc	1h 28m	4° 31'	-11.6	29'	87%	140° E	+G		Mars
27th	Cet	2h 12m	9° 19'	-11.9	29'	92%	149° E	+G		Uranus
28th	Ari	2h 57m	13° 47'	-12.2	30'	97%	159° E	FM		Uranus
29th	Tau	3h 44m	17° 45'	-12.5	30'	99%	170° E	FM		Pleiades, Hyades
30th	Tau	4h 33m	21° 1'	-12.7	30'	100%	179° W	FM		Pleiades, Hyades, Aldebaran

Mercury and Venus

Mercury
25th

Venus
25th

Mercury

Date	Con.	R.A.	Dec.	Mag.	Diam.	Ill.	Elon.	Vis.	Rat.		Close To
21st	Lib	14h 48m	-14° 30'	-0.7	5"	86%	15° W	AM	**		
23rd	Lib	15h 0m	-15° 35'	-0.7	5"	89%	14° W	NV	N/A		
25th	Lib	15h 12m	-16° 38'	-0.7	5"	91%	14° W	NV	N/A		
27th	Lib	15h 24m	-17° 39'	-0.7	5"	93%	13° W	NV	N/A		
29th	Lib	15h 36m	-18° 38'	-0.7	5"	95%	12° W	NV	N/A		

Venus

Date	Con.	R.A.	Dec.	Mag.	Diam.	Ill.	Elon.	Vis.	Rat.	Close To
21st	Vir	13h 53m	-9° 48'	-4.0	12"	87%	29° W	AM	**	Spica
23rd	Vir	14h 3m	-10° 40'	-4.0	12"	87%	29° W	AM	**	Spica
25th	Vir	14h 12m	-11° 31'	-4.0	12"	87%	28° W	AM	**	
27th	Vir	14h 22m	-12° 21'	-4.0	12"	88%	28° W	AM	**	
29th	Lib	14h 31m	-13° 11'	-4.0	12"	88%	28° W	AM	*	

Mars and the Outer Planets

Mars
25th

Jupiter
25th

Saturn
25th

Mars

Date	Con.	R.A.	Dec.	Mag.	Diam.	Ill.	Elon.	Vis.	Rat.	Close To
21st	Psc	0h 58m	5° 41'	-1.5	16"	94%	137° E	PM	****	
25th	Psc	0h 59m	6° 3'	-1.3	16"	93%	133° E	PM	****	Moon
30th	Psc	1h 2m	6° 36'	-1.2	15"	92%	129° E	PM	****	

The Outer Planets

Planet	Date	Con.	R.A.	Dec.	Mag.	Diam.	Elon.	Vis.	Rat.	Close To
Jupiter	25th	Sgr	19h 48m	-21° 34'	-2.0	35"	55° E	PM	**	Saturn
Saturn	25th	Sgr	20h 0m	-20° 55'	0.6	16"	58° E	PM	**	
Uranus	25th	Ari	2h 22m	13° 38'	5.7	4"	154° E	PM	***	
Neptune	25th	Aqr	23h 18m	-5° 41'	7.9	2"	108° E	PM	***	

Highlights

Date	Time (UT)	Event
22nd	04:46	First Quarter Moon. (Evening sky.)
	N/A	The Alpha Monocerotid meteor shower is at its maximum. (ZHR: Variable.)
23rd	10:52	The just-past first quarter Moon is south of Neptune. (Evening sky.)
25th	19:21	The waxing gibbous Moon is south of Mars. (Evening sky.)
27th	15:36	The waxing gibbous Moon is south of Uranus. (Evening sky.)
29th	05:27	Neptune is stationary prior to resuming prograde motion. (Evening sky.)
	12:46	The nearly full Moon is south of the Pleiades star cluster. (Evening sky.)
30th	09:30	Full Moon. (Visible all night.)
	09:43	Penumbral lunar eclipse. Visible from Asia, the northern Atlantic, Australia, western Europe, North America, the Pacific and South America.
	12:47	The full Moon is north of the bright star Aldebaran. (Taurus, visible all night.)

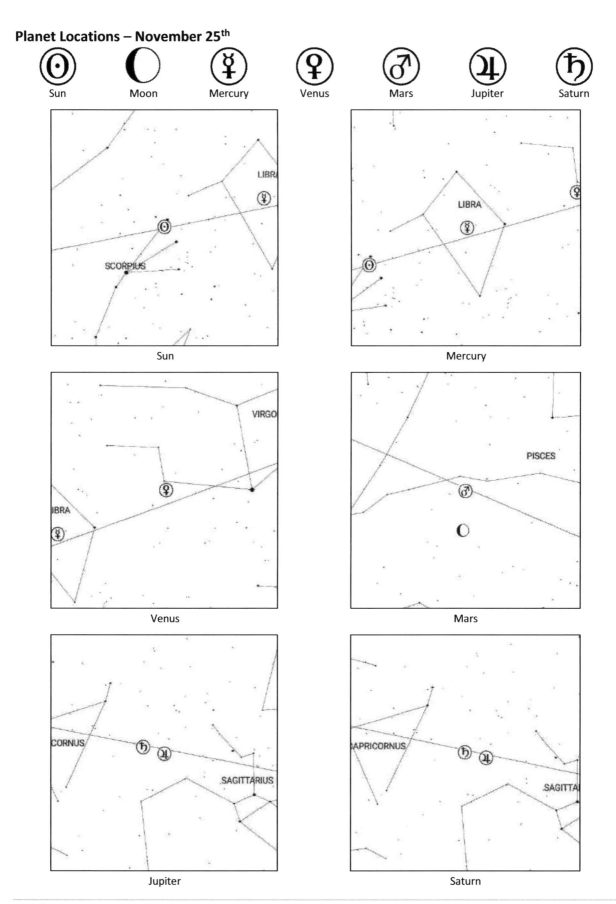

November 25ᵗʰ – Evening Sky

After passing Jupiter and Saturn and reaching first quarter, the Moon is now a waxing gibbous and can be seen to the lower right of Mars.

Despite rapidly fading, Mars itself is still shining brilliantly, and is brighter than all the stars in the night sky. Only the Moon and Jupiter are currently brighter.

This image depicts the sky looking east at about an hour after sunset.

November 25ᵗʰ - Evening Sky

The stars of winter are beginning to rise above the north-eastern horizon. You may be familiar with Taurus (Tau) the Bull, but its neighbor Auriga (Aur) the Charioteer is often overlooked.

Look out for Capella. It's the brightest yellow star in the sky and the sixth brightest overall. To our eyes, it appears as a single bright point, but there are actually four stars here: a pair of yellow giants and a pair of red dwarfs.

This image depicts the sky looking north-east at about two hours after sunset.

November 30ᵗʰ - Morning Sky

The Moon turns full today, but for observers in North America there's an added bonus: a penumbral lunar eclipse.

This happens when the Moon passes through the outer edge of the Earth's shadow, and although it's not as spectacular as a total eclipse, you should still see the Moon dim a little.

Mid-eclipse occurs at 4:43 a.m. Eastern Time. This image depicts the sky from Kansas City, KS, looking west at 03:43 a.m. Central Time.

December 1st to 10th, 2020

The Moon

  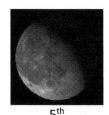  

1st	3rd	5th	7th	9th

Date	Con	R.A.	Dec	Mag	Diam	Ill.	Elon.	Phase	Close To
1st	Tau	5h 25m	23° 24'	-12.4	30'	99%	167° W	FM	
2nd	Gem	6h 19m	24° 41'	-12.1	30'	96%	154° W	FM	
3rd	Gem	7h 14m	24° 45'	-11.8	30'	91%	142° W	-G	
4th	Cnc	8h 9m	23° 33'	-11.5	31'	84%	129° W	-G	Praesepe
5th	Cnc	9h 4m	21° 6'	-11.2	31'	76%	116° W	-G	Praesepe
6th	Leo	9h 58m	17° 30'	-10.8	31'	66%	104° W	-G	Regulus
7th	Leo	10h 50m	12° 57'	-10.4	32'	56%	92° W	LQ	Regulus
8th	Vir	11h 42m	7° 39'	-9.9	32'	45%	80° W	LQ	
9th	Vir	12h 34m	1° 50'	-9.3	32'	34%	68° W	-Cr	
10th	Vir	13h 26m	-4° 11'	-8.6	33'	23%	56° W	-Cr	Spica

Mercury and Venus

Mercury
5th

Venus
5th

Mercury

Date	Con.	R.A.	Dec.	Mag.	Diam.	Ill.	Elon.	Vis.	Rat.	Close To
1st	Lib	15h 49m	-19° 33'	-0.8	5"	96%	11° W	NV	N/A	Antares
3rd	Sco	16h 2m	-20° 25'	-0.8	5"	97%	10° W	NV	N/A	Antares
5th	Sco	16h 15m	-21° 14'	-0.8	5"	98%	9° W	NV	N/A	Antares
7th	Oph	16h 28m	-21° 58'	-0.8	5"	98%	8° W	NV	N/A	Antares
9th	Oph	16h 41m	-22° 39'	-0.9	5"	99%	6° W	NV	N/A	Antares

Venus

Date	Con.	R.A.	Dec.	Mag.	Diam.	Ill.	Elon.	Vis.	Rat.	Close To
1st	Lib	14h 41m	-13° 59'	-4.0	12"	89%	28° W	AM	*	
3rd	Lib	14h 51m	-14° 45'	-4.0	12"	89%	27° W	AM	*	
5th	Lib	15h 1m	-15° 30'	-4.0	11"	90%	27° W	AM	*	
7th	Lib	15h 11m	-16° 14'	-4.0	11"	90%	27° W	AM	*	
9th	Lib	15h 21m	-16° 56'	-4.0	11"	90%	27° W	AM	*	

Mars and the Outer Planets

Mars
5th

Jupiter
5th

Saturn
5th

Mars

Date	Con.	R.A.	Dec.	Mag.	Diam.	Ill.	Elon.	Vis.	Rat.	Close To
1st	Psc	1h 3m	6° 43'	-1.1	15"	92%	128° E	PM	***	
5th	Psc	1h 6m	7° 13'	-1.0	14"	92%	124° E	PM	***	
10th	Psc	1h 11m	7° 53'	-0.8	13"	91%	120° E	PM	***	

The Outer Planets

Planet	Date	Con.	R.A.	Dec.	Mag.	Diam.	Elon.	Vis.	Rat.	Close To
Jupiter	5th	Sgr	19h 56m	-21° 12'	-2.0	34"	47° E	PM	**	Saturn
Saturn	5th	Sgr	20h 4m	-20° 45'	0.6	16"	49° E	PM	**	
Uranus	5th	Ari	2h 20m	13° 32'	5.7	4"	143° E	PM	***	
Neptune	5th	Aqr	23h 18m	-5° 41'	7.9	2"	97° E	PM	***	

Highlights

Date	Time (UT)	Event
5th	00:06	The waning gibbous Moon is north of the Praesepe star cluster. (Cancer, morning sky.)
6th	16:38	The waning gibbous Moon is north of the bright star Regulus. (Leo, morning sky.)
8th	00:37	Last Quarter Moon. (Morning sky.)
9th	N/A	The Monocerotid meteor shower is at its maximum. (ZHR: 2)
10th	12:46	The waning crescent Moon is north of the bright star Spica. (Virgo, morning sky.)

Planet Locations – December 5th

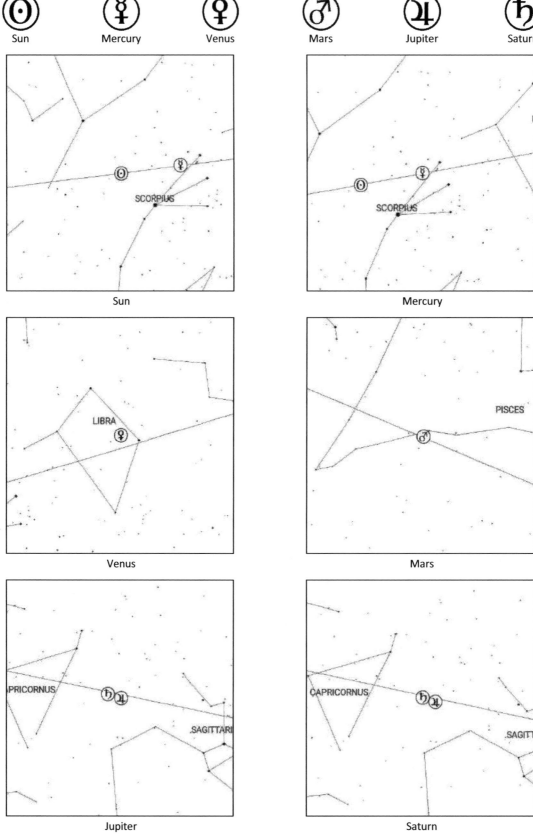

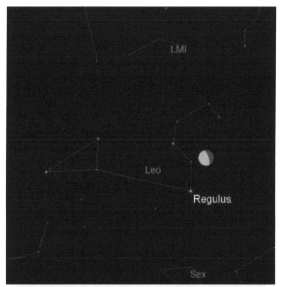

December 6th – Morning Sky

It's a quiet start to the month with little happening in terms of astronomical events.

That being said, if you're up early, you can catch the waning gibbous Moon close to the bright star Regulus in the pre-dawn sky.

This image depicts the sky looking south at about 75 minutes before sunrise.

December 8th – Morning Sky

The Moon reaches last quarter today and is now leaving Leo, the Lion, behind and approaching Virgo (Vir), the Virgin.

It will appear close to Spica, that constellation's brightest star, in a couple of days.

This image depicts the sky looking south-east about 90 minutes before sunrise.

December 10th – Morning Sky

Venus is an early riser and can be seen low over the south-eastern horizon before the dawn.

This morning it's joined by the waning crescent Moon. It's now close to Spica and will appear near to the planet in just another two days.

This image depicts the sky looking south-east about 90 minutes before sunrise.

December 11th to 20th, 2020

The Moon

| | 11th | | 13th | | 15th | | 17th | | 19th |

Date	Con	R.A.	Dec	Mag	Diam	Ill.	Elon.	Phase	Close To
11th	Vir	14h 20m	-10° 6'	-7.8	33'	14%	44° W	-Cr	
12th	Lib	15h 16m	-15° 30'	-6.8	33'	7%	31° W	NM	Venus
13th	Sco	16h 16m	-19° 59'	-5.6	33'	2%	17° W	NM	Venus, Antares
14th	Oph	17h 18m	-23° 9'	-4.1	33'	0%	3° W	NM	Mercury
15th	Sgr	18h 22m	-24° 43'	-5.1	33'	1%	12° E	NM	
16th	Sgr	19h 25m	-24° 37'	-6.3	32'	4%	27° E	NM	Jupiter, Saturn
17th	Cap	20h 25m	-22° 58'	-7.4	32'	10%	41° E	NM	Jupiter, Saturn
18th	Cap	21h 21m	-20° 4'	-8.2	31'	18%	54° E	+Cr	
19th	Aqr	22h 14m	-16° 14'	-8.9	31'	26%	66° E	+Cr	
20th	Aqr	23h 2m	-11° 47'	-9.4	30'	36%	77° E	+Cr	Neptune

Mercury and Venus

Mercury
15th

Venus
15th

Mercury

Date	Con.	R.A.	Dec.	Mag.	Diam.	Ill.	Elon.	Vis.	Rat.	Close To
11th	Oph	16h 55m	-23° 14'	-0.9	5"	99%	5° W	NV	N/A	Antares
13th	Oph	17h 8m	-23° 45'	-1.0	5"	100%	4° W	NV	N/A	Antares
15th	Oph	17h 22m	-24° 12'	-1.1	5"	100%	3° W	NV	N/A	
17th	Oph	17h 35m	-24° 33'	-1.1	5"	100%	2° W	NV	N/A	
19th	Sgr	17h 49m	-24° 49'	-1.2	5"	100%	0° W	NV	N/A	

Venus

Date	Con.	R.A.	Dec.	Mag.	Diam.	Ill.	Elon.	Vis.	Rat.	Close To
11th	Lib	15h 31m	-17° 36'	-4.0	11"	91%	26° W	AM	*	
13th	Lib	15h 41m	-18° 15'	-4.0	11"	91%	26° W	AM	*	Moon
15th	Lib	15h 51m	-18° 51'	-4.0	11"	91%	26° W	AM	*	Antares
17th	Lib	16h 1m	-19° 26'	-3.9	11"	92%	25° W	AM	*	Antares
19th	Sco	16h 12m	-19° 58'	-3.9	11"	92%	25° W	AM	*	Antares

Mars and the Outer Planets

Mars
15th

Jupiter
15th

Saturn
15th

Mars

Date	Con.	R.A.	Dec.	Mag.	Diam.	Ill.	Elon.	Vis.	Rat.	Close To
11th	Psc	1h 12m	8° 2'	-0.8	13"	91%	119° E	PM	***	
15th	Psc	1h 16m	8° 38'	-0.7	12"	90%	116° E	PM	***	
20th	Psc	1h 23m	9° 25'	-0.5	12"	90%	112° E	PM	***	

The Outer Planets

Planet	Date	Con.	R.A.	Dec.	Mag.	Diam.	Elon.	Vis.	Rat.	Close To
Jupiter	15th	Sgr	20h 5m	-20° 48'	-2.0	34"	38° E	PM	**	Saturn
Saturn	15th	Sgr	20h 8m	-20° 33'	0.6	15"	39° E	PM	**	
Uranus	15th	Ari	2h 19m	13° 27'	5.7	4"	131° E	PM	***	
Neptune	15th	Aqr	23h 18m	-5° 39'	7.9	2"	86° E	PM	***	

Highlights

Date	Time (UT)	Event
11th	N/A	Good opportunity to see Earthshine on the waning crescent Moon. (Morning sky.)
12th	21:05	The waning crescent Moon is north of Venus. (Morning sky.)
14th	16:15	Total solar eclipse. Visible from Antarctica, the southern Atlantic, the southern Pacific and South America.
	16:17	New Moon. (Not visible.)
	N/A	The Geminid meteor shower is at its maximum. (ZHR: 120)
17th	04:00	The waxing crescent Moon is south of Jupiter. (Evening sky.)
	04:39	The waxing crescent Moon is south of Saturn. (Evening sky.)
18th	N/A	Good opportunity to see Earthshine on the waxing crescent Moon. (Evening sky.)
20th	03:09	Mercury is at superior conjunction with the Sun. (Not visible.)
	21:18	The nearly first quarter Moon is south of Neptune. (Evening sky.)
	N/A	The Leonis Minorid meteor shower is at its maximum. (ZHR: 5)

Planet Locations – December 15th

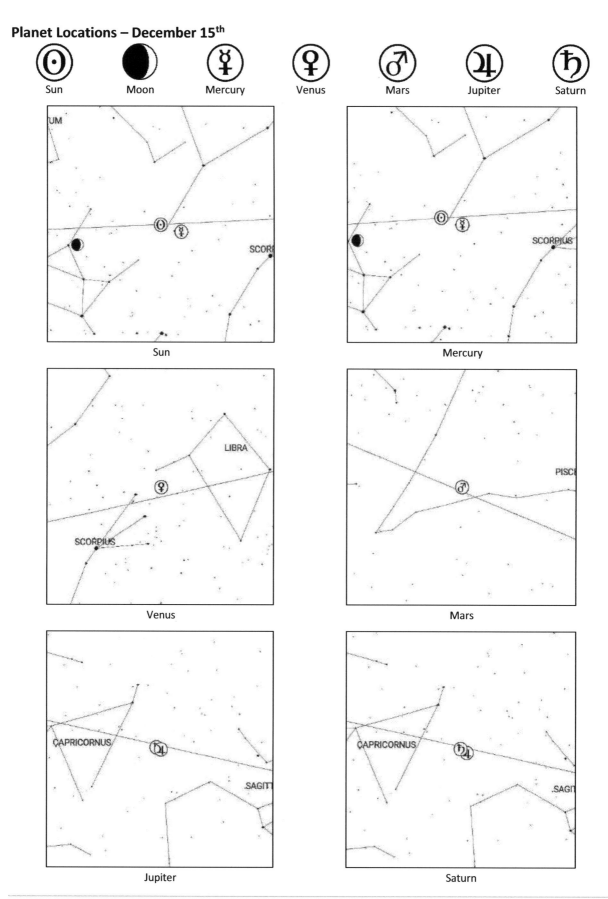

December 12th – Morning Sky

The waning crescent Moon and Venus form a very nice pair in the pre-dawn sky this morning.

They're both among the stars of Libra (Lib), the Scales. This constellation is one of the faintest in the zodiac and is the only zodiac sign that represents an inanimate object.

This image depicts the sky looking south-east at about 45 minutes before sunrise.

December 13th to 14th - Evening & Morning Sky

The Geminid meteor shower reaches its maximum today and promises to put on a good show for observers. With the Moon turning new, there'll be no interference from lunar light.

Theoretically, the Geminids could produce up to 120 shooting stars an hour under ideal conditions, making this a shower worth looking out for.

This image depicts the sky looking east at about 8:00 p.m.

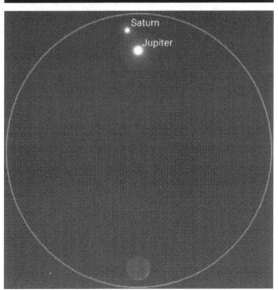

December 16th – Evening Sky

There's a thin sliver of a crescent Moon in the evening twilight and tonight it's close to both Jupiter and Saturn.

Jupiter is rapidly catching up to its slower sibling and the pair will be at their closest on the 21st. In the meantime, they look great in binoculars with the Moon just a little below them.

This image depicts the view through 10x50 binoculars looking south-west at about an hour after sunset.

December 21st to 31st, 2020

The Moon

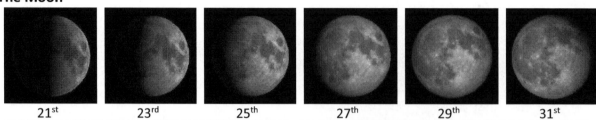

| 21st | 23rd | 25th | 27th | 29th | 31st |

Date	Con	R.A.	Dec	Mag	Diam	Ill.	Elon.	Phase	Close To
21st	Aqr	23h 48m	-6° 57'	-9.9	30'	45%	87° E	FQ	Neptune
22nd	Cet	0h 32m	-1° 58'	-10.3	30'	55%	97° E	FQ	
23rd	Psc	1h 16m	3° 2'	-10.7	30'	64%	107° E	FQ	Mars
24th	Psc	1h 59m	7° 54'	-11.0	30'	73%	116° E	+G	Mars, Uranus
25th	Ari	2h 43m	12° 28'	-11.4	30'	81%	126° E	+G	Uranus
26th	Tau	3h 29m	16° 36'	-11.7	30'	88%	137° E	+G	Pleiades
27th	Tau	4h 18m	20° 7'	-11.9	30'	93%	148° E	+G	Pleiades, Hyades, Aldebaran
28th	Tau	5h 9m	22° 47'	-12.2	30'	97%	160° E	FM	Hyades, Aldebaran
29th	Gem	6h 3m	24° 26'	-12.5	30'	100%	172° E	FM	
30th	Gem	6h 59m	24° 51'	-12.6	31'	100%	175° W	FM	
31st	Gem	7h 55m	23° 58'	-12.3	31'	98%	162° W	FM	Praesepe

Mercury and Venus

Mercury
25th

Venus
25th

Mercury

Date	Con.	R.A.	Dec.	Mag.	Diam.	Ill.	Elon.	Vis.	Rat.	Close To
21st	Sgr	18h 3m	-24° 60'	-1.2	5"	100%	1° E	NV	N/A	
23rd	Sgr	18h 17m	-25° 5'	-1.1	5"	100%	2° E	NV	N/A	
25th	Sgr	18h 31m	-25° 4'	-1.1	5"	100%	3° E	NV	N/A	
27th	Sgr	18h 46m	-24° 58'	-1.0	5"	99%	5° E	NV	N/A	
29th	Sgr	19h 0m	-24° 45'	-1.0	5"	99%	6° E	NV	N/A	
31st	Sgr	19h 14m	-24° 27'	-1.0	5"	98%	7° E	NV	N/A	

Venus

Date	Con.	R.A.	Dec.	Mag.	Diam.	Ill.	Elon.	Vis.	Rat.	Close To
21st	Sco	16h 22m	-20° 28'	-3.9	11"	92%	24° W	AM	*	Antares
23rd	Oph	16h 33m	-20° 56'	-3.9	11"	93%	24° W	AM	*	Antares
25th	Oph	16h 43m	-21° 21'	-3.9	11"	93%	24° W	AM	*	Antares
27th	Oph	16h 54m	-21° 44'	-3.9	11"	93%	23° W	AM	*	Antares
29th	Oph	17h 5m	-22° 4'	-3.9	11"	94%	23° W	AM	*	Antares
31st	Oph	17h 15m	-22° 22'	-3.9	11"	94%	22° W	AM	*	

Mars and the Outer Planets

Mars
25th

Jupiter
25th

Saturn
25th

Mars

Date	Con.	R.A.	Dec.	Mag.	Diam.	Ill.	Elon.	Vis.	Rat.	Close To
21st	Psc	1h 24m	9° 34'	-0.5	12"	90%	111° E	PM	***	
25th	Psc	1h 30m	10° 14'	-0.4	11"	89%	108° E	PM	***	
31st	Psc	1h 39m	11° 15'	-0.3	10"	89%	104° E	PM	***	

The Outer Planets

Planet	Date	Con.	R.A.	Dec.	Mag.	Diam.	Elon.	Vis.	Rat.	Close To
Jupiter	25th	Cap	20h 14m	-20° 20'	-2.0	33"	29° E	PM	*	Saturn
Saturn	25th	Cap	20h 12m	-20° 20'	0.6	15"	29° E	PM	**	
Uranus	25th	Ari	2h 18m	13° 23'	5.7	4"	120° E	PM	***	Moon
Neptune	25th	Aqr	23h 18m	-5° 36'	7.9	2"	75° E	PM	**	

Highlights

Date	Time (UT)	Event
21st	10:04	Winter Solstice.
	13:46	Jupiter is 0.1° south of Saturn. (Evening sky.)
	23:42	First Quarter Moon. (Evening sky.)
22nd	18:51	Venus is 5.7° north of the bright star Antares. (Scorpius, morning sky.)
23rd	18:09	The waxing gibbous Moon is south of Mars. (Evening sky.)
	N/A	The Ursid meteor shower is at its maximum. (ZHR: 10)
24th	23:51	The waxing gibbous Moon is south of Uranus. (Evening sky.)
26th	20:11	The waxing gibbous Moon is south of the Pleiades star cluster. (Taurus, evening sky.)
27th	19:28	The waxing gibbous Moon is north of the bright star Aldebaran. (Taurus, evening sky.)
30th	03:29	Full Moon. (Visible all night.)

Planet Locations – December 25th

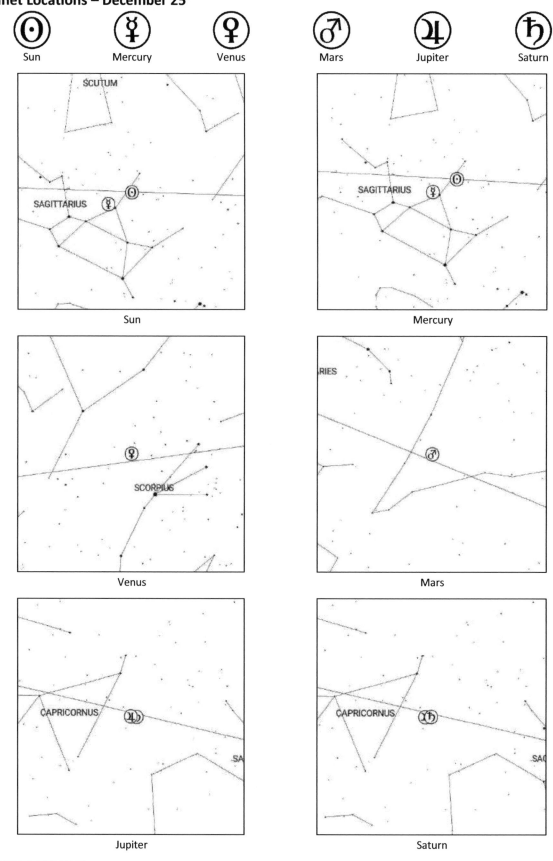

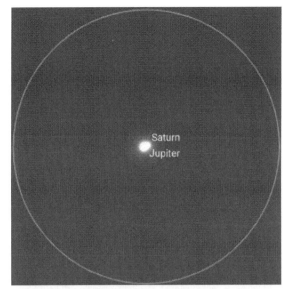

December 21st – Evening Sky

Jupiter has finally caught up to Saturn in the evening sky. This rare event only happens once every twenty years and the pair won't be this close again until 2080!

This year there's just a tenth of a degree behind them, which is roughly a third of the width of the full Moon. Jupiter, the brighter of the two, appears just to the left of Saturn.

This image depicts the view through 10x50 binoculars looking south-west about an hour after sunset.

December 23rd – Morning Sky

Venus continues to slip lower in the pre-dawn sky and provides us with a challenge this morning.

The planet is easy enough to see, but can you spot Antares, the brightest star in Scorpius (Sco), the Scorpion? You'll need a clear, unobstructed view of the horizon to see it!

This image depicts the sky looking south-east about 45 minutes before sunrise.

December 23rd – Evening Sky

Our last night sky sight involves the Moon and Mars. The planet reached opposition a few months ago and despite continuing to dim, can still be seen shining brightly in the evening sky.

Tonight you'll find the waxing gibbous Moon almost directly below it. Pisces (Psc) the Fish is a faint constellation, but can you see any of its stars nearby?

This image depicts the sky looking south about two hours after sunset.

Glossary

Aphelion

The point at which an object is farthest from the Sun. (See also *perihelion*.)

Apogee

The point at which an object is farthest from the Earth. (See also *perigee*.)

Apparent Diameter

Apparent Diameter is the size an object appears in the sky and is measured in degrees, arc-minutes and arc-seconds. If you were to stand facing due north and slowly turn toward east, south, west and then north again, you would be turning 360° (degrees.)

You might therefore think that you can see 360° of sky overhead – in fact, you can only see 180° because the sky is only the visible half a sphere and since you can't see the entire sphere (the ground is in the way,) you can't look 360° in every direction at the sky.

If, however, you were to face due north and look directly overhead at the zenith, this would be 90°. Look down from the zenith to the southern horizon and you would see another 90°, making 180° total. (Incidentally, how high an object appears in the sky is called its *altitude*, but it isn't necessary to know that to use this book.)

To put this into perspective, the Sun and Moon both appear to be about half a degree in diameter but because these are the largest astronomical objects in the sky and everything else appears to be smaller, we need a more convenient (and accurate) measurement.

A degree then is broken up into sixty arc-minutes. Therefore, because the Sun and Moon both appear to be about half a degree, we say their apparent diameter is 30' (arc-minutes.)

The planets and asteroids are even smaller, so we break each arc-minute up into sixty again, thereby creating arc-seconds. A planets' apparent diameter will greatly depend on its actual size and its distance from Earth.

For example, although the planet Venus is slightly smaller than the Earth, it is also the closest world to our own. So at its closest (inferior conjunction) it can have an apparent diameter of 66.01" – in other words, 1' 01" (one arc-minute and one arc-second.) Despite Jupiter being large enough to swallow all the other planets within it, it is much further away – therefore, at its best, it is only able to reach 50.12" (fifty arc-seconds). Neptune is the fourth largest planet but is also the most distant and barely manages to reach 2.37" (arc-seconds.)

Even through a small telescope observers can easily see all of the planets as discs; however, how large the planet appears and the details seen will vary greatly, depending upon the equipment used and the apparent diameter of the planet itself. The dark bands of Jupiter's' atmosphere are easily visible in almost any sized 'scope, whereas Uranus and Neptune typically only show tiny discs in a small to medium sized telescope. Under low power those distant worlds can easily be mistaken for stars.

Only the apparent diameters of the Moon and planets are noted in this book; the dwarf planets Pluto and Ceres, along with the asteroids Juno, Pallas and Vesta and all the bright stars mentioned, only appear as points of light through amateur instruments and their apparent diameter is therefore negligible.

To help put this into perspective, the image below depicts the average apparent size of the planets in comparison with one another.

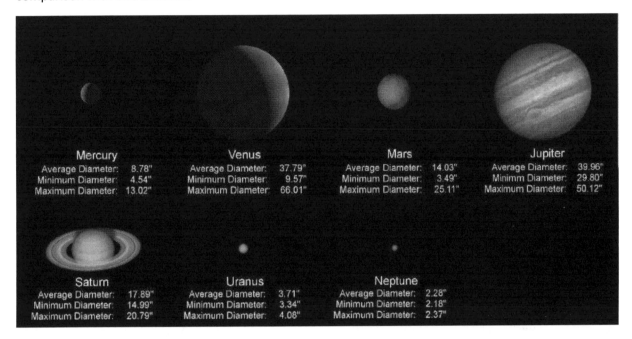

Mercury		Venus		Mars		Jupiter	
Average Diameter:	8.78"	Average Diameter:	37.79"	Average Diameter:	14.03"	Average Diameter:	39.96"
Minimum Diameter:	4.54"	Minimum Diameter:	9.57"	Minimum Diameter:	3.49"	Minimm Diameter:	29.80"
Maximum Diameter:	13.02"	Maximum Diameter:	66.01"	Maximum Diameter:	25.11"	Maximum Diameter:	50.12"

Saturn		Uranus		Neptune	
Average Diameter:	17.89"	Average Diameter:	3.71"	Average Diameter:	2.28"
Minimum Diameter:	14.99"	Minimum Diameter:	3.34"	Minimum Diameter:	2.18"
Maximum Diameter:	20.79"	Maximum Diameter:	4.08"	Maximum Diameter:	2.37"

Asterism

An asterism is a recognizable pattern of stars within a constellation. For example, the "backwards question mark" depicting the head of Leo the Lion, or the seven stars of the Big Dipper (aka, the Plough) in the much-larger constellation of Ursa Major, the Great Bear.

AU – Astronomical Unit

An astronomical unit is basically the mean distance of the Earth to the Sun and is the standard measurement of distance within the solar system. One astronomical unit is equivalent to almost 150 million kilometers (specifically, 149,597,870 km) or roughly 92.96 million miles.

Conjunction

Astronomically, this is a fairly vague term. It basically means any situation when two bodies appear close to one another in the sky. However, there is no officially recognized separation limit that would clearly define when a conjunction is taking place. An object in conjunction with the Sun is never visible because the light from the Sun is too over-powering and the object will be lost in the glare. (See also *inferior conjunction, opposition* and *superior conjunction.*)

Culminate

Culmination occurs when an object is at its highest point in the sky. For most objects, this means it is due south, but this will depend on the object and your latitude. For example, some stars or objects may be directly overhead when they culminate. (In fact, if you faced due south and followed an invisible line between due south and due north, your gaze would pass overhead and any object culminating could also be in sight.)

Earthshine

Earthshine is when the Moon is a crescent but you can see the "dark side" of the Moon too – so you can see the whole Moon in the sky. This happens when light is reflected from the daylight side of the Earth and illuminates the unlit portion of the Moons' surface. It can make for a beautiful sight in the twilight, especially when the Moon is close to a bright star or planet.

Left: An example of Earthshine on the waxing crescent Moon. Light is reflected from the Earth, causing the "dark side" of the Moon to be visible. In this image, the unlit portion of the Moon has been lightened to be more apparent. Image by Steve Jurvetson and used under the Creative Commons Attribution 2.0 Generic license.

Ecliptic

The approximate path the Sun, Moon and planets appear to follow across the sky. The ecliptic crosses the traditional twelve signs of the zodiac as well as briefly passing through other constellations, such as Ophiuchus and Orion. The ecliptic is depicted as a pale blue line in the images used throughout this book.

Elongation

Elongation is how far to the east or west an object appears in relation to another object (most usually in relation to the Sun.) If Mercury or Venus is at eastern elongation, it will appear in the evening sky. If Mercury or Venus is at western elongation, it will appear in the pre-dawn sky. (It is worth noting that there is no guarantee the planet will be visible – it will also depend upon the time of year and the observers' latitude.)

Gibbous

The Moon is said to be gibbous between the half phases (first and last quarter) and full Moon. It's hard to describe the shape – it's not a half Moon, but it's not completely circular either. The inner planets Mercury and Venus can also show a gibbous phase. (See also *illumination*.)

Globular Star Cluster

A globular star cluster is, quite literally and simply, a sphere of thousands of stars. Globular clusters appear as faint, misty balls of grey light against the night, and although they all require at least a pair of binoculars to be seen, many can be resolved into their individual stars through a telescope. The best (and most famous) example in the northern hemisphere is M13, the Great Hercules Cluster (see image below and also *open star cluster*.)

Left: M13, the Great Hercules Cluster. Photo taken by the author using Slooh.

Illumination

Simply how much of an object's visible surface is lit. For example, when the Moon is new, none of the lit surface is visible, so the Moon is 0% illuminated. At half phase (first or last quarter) half the lit surface is visible, so the Moon is 50% illuminated. At full Moon, the entire lit surface is visible, so the Moon is 100% illuminated. The planets Mercury, Venus and Mars can also show phases (Mars, being an outer planet, is more limited) and so their illumination will also change over time. (See also *gibbous*.)

Inferior Conjunction

Inferior conjunction occurs when either Mercury or Venus are directly between the Earth and the Sun. Because the planet will appear so close to the Sun in the sky, it will not be visible from Earth. Mercury and Venus are the only two planets that can go through inferior conjunction because only these two worlds orbit closer to the Sun than the Earth. (See also *conjunction* and *superior conjunction.*)

Magnitude

An object's magnitude is simply a measurement of its brightness. The ancient Greeks created a system where the brightest stars were given a magnitude of 1 and the faintest were magnitude 6. Since that time, astronomers have refined the system and increased its accuracy, but as a result, the magnitude range has increased dramatically.

For example, there are some objects that are brighter than zero and therefore have a negative magnitude. Sirius, the brightest star in the sky, has a magnitude of -1.47. All of the naked eye planets – Mercury, Venus, Mars, Jupiter and Saturn – can all have negative magnitudes. The other two planets, Uranus and Neptune, dwarf planets and asteroids are all more than magnitude five.

(The bright star Vega, in the constellation Lyra, is used as the standard reference point. It has a magnitude of 0.0.)

The naked eye can, theoretically, see objects up to magnitude six, but this greatly depends upon the observer's vision and the conditions of the night sky. Most people can see up to around magnitude five under clear, dark, rural skies. Light pollution is so bad in many towns and cities that, even in the suburbs, it is often difficult to see anything fainter than magnitude 3 or 4 at best.

However, the Moon, Mercury, Venus, Mars, Jupiter and Saturn should easily be visible to anyone, anywhere, assuming that the object is not too close to the Sun. A planets' magnitude will vary, depending upon how close it is to the Earth, how large it appears in the sky and how much of its lit surface is visible. (See also *apparent diameter* and *illumination*.)

Constellations can be problematic, depending upon the brightness of the stars that form the constellation itself.

Meteors are best observed from rural skies but the bright stars mentioned, as well as the Pleiades and Hyades star clusters can be seen from the suburbs.

Almost everything else is best observed under rural skies and/or with binoculars or a telescope.

The magnitude ranges of the solar system objects mentioned in this book are detailed below. (The Sun is, on average, about magnitude -26.74)

Object	Minimum Magnitude (Faintest)	Maximum Magnitude (Brightest)
Moon	-2.5 (New)	-12.9 (Full)
Mercury	5.73	-2.45
Venus	-3.82	-4.89
Mars	1.84	-2.91
Jupiter	-1.61	-2.94
Saturn	1.47	-0.49
Uranus	5.95	5.32
Neptune	8.02	7.78
Pluto	16.3	13.65
Dwarf planet Ceres	9.34	6.64
Asteroid 2 Pallas	10.65	6.49
Asteroid 3 Juno	11.55	7.4
Asteroid 4 Vesta	8.48	5.1

Occultation

When one object completely covers another. Many occultations involve the Moon occulting a star or, sometimes, a planet, but on occasion, a planet may be seen to occult a star. On very rare occasions, one planet may occult another.

Open Star Cluster

An open star cluster is one where the stars appear to be loosely scattered against the night. Unlike globular clusters, their member stars are usually quite young and only number a couple of hundred at most. Several open clusters can be seen with the naked eye (for example, M44 - the Praesepe - in Cancer and the Hyades, which forms the V shaped asterism in the constellation Taurus.) The most famous example of an open cluster is M45, the Pleiades, a naked eye cluster (also in Taurus) that is easily seen throughout the winter. (See image below and also *globular star cluster*.)

Left: M45, the famous Pleiades open star cluster in Taurus. Easily visible with the naked eye throughout the winter, this image shows the deep blue nebulosity that surround the stars. This nebulosity is the remains of the cloud that gave birth to the stars themselves, but unfortunately this is not visible to the vast majority of observers. Photo by the author using Slooh.

Opposition

An object is said to be at opposition when it is directly opposite the Sun in the sky. On that date, it is visible throughout the night as it will rise at sunset, culminate at midnight and set at sunrise. For that reason, this is the best opportunity to observe that object. (See also *conjunction*.)

Perihelion

The point at which an object is closest to the Sun. (See also *aphelion*.)

Perigee

The point at which an object is closest to the Earth. (See also *apogee*.)

Prograde Motion

Prograde motion is when a body appears to move forwards through the constellations from west to east. It's the normal motion of the Sun, Moon, planets and asteroids across the sky. (See also *retrograde motion*.)

Retrograde Motion

Retrograde motion is when a body appears to move *backwards* through the constellations, from east to west. For the inferior planets Mercury and Venus, this happens for a time after greatest eastern elongation (when the planet appears in the evening sky) and before greatest western elongation (when it appears in the pre-dawn sky) as the planet catches up to and then passes the Earth in its orbit. For all the other planets and asteroids, it happens for a time before and after opposition when the Earth catches up to that world and then passes it. This YouTube video from 2009 does a good job of graphically depicting how this happens. (See also *prograde motion*.)

Superior Conjunction

Superior conjunction occurs when either Mercury or Venus are on the opposite side of the Sun from the Earth. For example, if Mercury is at superior conjunction, the Sun would be directly between the Earth and the Mercury. Like *inferior conjunction*, the planet appears very close to the Sun in the sky and is not visible from Earth.

Again, like *inferior conjunction,* Mercury and Venus are the only two planets that can go through inferior conjunction because only these two worlds orbit closer to the Sun than the Earth. (See also *conjunction* and *inferior conjunction.*)

Universal Time

Universal Time is the standard method of notating when an astronomical event takes place. It is based upon Greenwich Mean Time and requires adjustment for other time zones:

Greenwich Mean Time – no change. (Summer Time – add one hour.)

Eastern Time – deduct five hours. (Summer Time – deduct four hours.)

Central Time – deduct six hours. (Summer Time – deduct five hours.)

Mountain Time – deduct seven hours. (Summer Time – deduct six hours.)

Pacific Time – deduct eight hours. (Summer Time – deduct seven hours.)

A useful website that will convert Universal Time to other time zones can be found at the following address: http://www.worldtimeserver.com/convert_time_in_UTC.aspx

Bear in mind that if an event takes place during the day at your location, it may still be visible in the evening or pre-dawn sky. For example, two planets may be at their closest at 1pm local time but because they don't move quickly, they'll still be very close together during the night. The only exception is the Moon – it *does* move relatively quickly, but may still appear fairly close to the object when it next becomes visible. It just won't be as close as it was at the time listed in the book.

Waning

The Moon is said to be "waning" between the full and new Moon. When the Moon is full, the Earth lies between the Moon and the Sun and the lit surface is completely visible to us. When the Moon wanes, the visible, lit portion of the Moon appears to decrease until it is completely invisible at new Moon. A waning Moon is best seen in the pre-dawn sky. (See also *waxing*.)

Waxing

The Moon is said to be "waxing" between the new and full Moon. When the Moon is new, it lies between the Earth and the Sun and the lit surface is not visible to us. When the Moon waxes, the visible, lit portion of the Moon appears to be increasing until it is completely lit at full Moon. A waxing Moon is best seen in the evening sky. (See also *waning*.)

Zenith

The point directly overhead in the sky. (See also *zenith hourly rate*.)

Zenith Hourly Rate

The number of meteors an observer can expect to see each hour at the zenith (directly overhead) on the shower's peak date. It's worth remembering that meteor showers can be somewhat unpredictable and, hence, the maximum zenith hourly rate is only an estimate at best. (See also *zenith*.)

Also by the Author

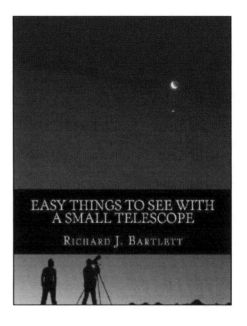

Signposts to the Stars

Easy Things to See With a Small Telescope

Aimed at absolute beginners, this book will help you to locate and learn the constellations using the brightest stars of Ursa Major and Orion as signposts.

More than that, the book also details:
*Key astronomical terms and phrases
*The brightest stars and constellations for each season
*The myths and legends of the stars
*Fascinating stars, star clusters, nebulae and galaxies, many of which can be seen with just your eyes or binoculars
*An introduction to the planets, comets and meteor showers

If you've ever stopped and stared at the stars but didn't know where to begin, these signposts will get you started on your journey!

Specifically written with the beginner in mind, this book highlights over sixty objects easily found and observed in the night sky. Objects such as:
* Stunning multiple stars
* Star clusters
* Nebulae
* And the Andromeda Galaxy!

Each object has its own page which includes a map, a view of the area through your finderscope and a depiction of the object through the eyepiece.

There's also a realistic description of every object based upon the author's own notes written over years of observations.

Additionally, there are useful tips and tricks designed to make your start in astronomy easier and pages to record your observations.

If you're new to astronomy and own a small telescope, this book is an invaluable introduction to the night sky.

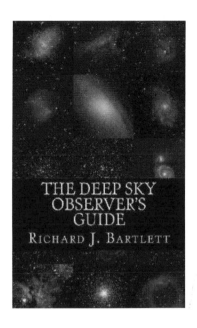

The Easy Guide to the Night Sky

The Deep Sky Observer's Guide

Written for the amateur astronomer who wants to discover more in the night sky, this book explores the constellations and reveals many of the highlights visible with just your eyes or binoculars.

The highlights include:
* The myths and legends associated with the stars
* Bright stars and multiple stars
* Star clusters
* Nebulae
* Galaxies

Each constellation has its own star chart and almost all are accompanied by graphics depicting the highlights and binocular views of the best objects.

Whether you're new to astronomy or are an experienced stargazer simply looking to learn more about the constellations, this book is an invaluable guide to the night sky and the stars to be found there.

The Deep Sky Observer's Guide offers you the night sky at your fingertips. As an amateur astronomer, you want to know what's up tonight and you don't always have the time to plan ahead. The Deep Sky Observer's Guide can solve this problem in a conveniently sized paperback that easily fits in your back pocket. Take it outside and let the guide suggest any one of over 1,300 deep sky objects, all visible with a small telescope and many accessible via binoculars.

* Multiple stars with 2" or more of separation
* Open clusters up to magnitude 9
* Nebulae up to magnitude 10
* Globular clusters up to magnitude 10
* Planetary nebulae up to magnitude 12
* Galaxies up to magnitude 12
* Includes lists of deep sky objects for the entire sky with R.A. and declination for each and accompanying images for many

Whether you use a GoTo or prefer to star hop, no matter where you live in the world and no matter what time of year or night, the Deep Sky Observer's Guide is the indispensable companion for every adventure among the stars.

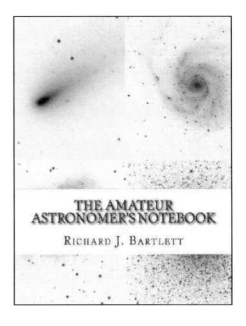

The Wonder of it All

The Amateur Astronomer's Notebook

From our home here on Earth, past the Sun, Moon and planets, this is a journey out to the stars and beyond.

A journey of discovery that shows us the beauty and wonder of the cosmos and our special and unique place within it.

Written by an amateur astronomer with a life-long love of the stars, The Wonder of It All will open your child's eyes to the universe and includes notes for parents to help develop an interest in astronomy.

The Amateur Astronomer's Notebook is the perfect way to log your observations of the Moon, stars, planets and deep sky objects.

With an additional appendix with hundreds of suggested deep sky objects, this 8.5" by 11" notebook allows you to record everything you need for 150 observing sessions under the stars:
*Date
*Time
*Lunar Phase
*Limiting Magnitude
*Transparency
*Seeing
*Equipment
*Eyepieces
*Additional Notes
*Pre-drawn circles to sketch your observations
*Plenty of room to record your notes and impressions

Whether you're an experienced astronomer or just beginning to discover the universe around us, you'll find the notebook to be an invaluable tool and record of your exploration of the cosmos.

Made in the USA
San Bernardino, CA
19 December 2019